Computational Fluid Dynamics for Engineers

Computational fluid dynamics (CFD) has become an indispensable tool for many engineers. This book gives an introduction to CFD simulations of turbulence, mixing, reaction, combustion and multiphase flows. The emphasis on understanding the physics of these flows helps the engineer to select appropriate models with which to obtain reliable simulations. Besides presenting the equations involved, the basics and limitations of the models are explained and discussed. The book, combined with tutorials, project and Power-Point lecture notes (all available for download), forms a complete course. The reader is given hands-on experience of drawing, meshing and simulation. The tutorials cover flow and reactions inside a porous catalyst, combustion in turbulent non-premixed flow and multiphase simulation of evaporating sprays. The project deals with the design of an industrial-scale selective catalytic reduction process and allows the reader to explore various design improvements and apply best practice guidelines in the CFD simulations.

Bengt Andersson is a Professor in Chemical Engineering at Chalmers University, Sweden. His research has focused on experimental studies and modelling of mass and heat transfer in various chemical reactors ranging from automotive catalysis to three-phase flow in chemical reactors.

Ronnie Andersson is an Assistant Professor in Chemical Engineering at Chalmers University. He obtained his PhD at Chalmers in 2005 and from 2005 until 2010 he worked as a consultant at Epsilon HighTech as a specialist in CFD simulations of combustion and multiphase flows. His research projects involve physical modelling, fluid-dynamic simulations and experimental methods.

Love Håkansson works as a consultant at Engineering Data Resources – EDR in Oslo, Norway. His research has been in mass transfer in turbulent boundary layers. He is now working on simulations of single-phase and multiphase flows.

Mikael Mortensen obtained his PhD at Chalmers University in 2005 in turbulent mixing with chemical reactions. After two years as a post doc at the University of Sydney, he is now working with fluid dynamics at the Norwegian Defence Research Establishment in Lillehammer, Norway.

Rahman Sudiyo is a Lecturer at the University of Gadjah Mada in Yogyakarta, Indonesia. He received his PhD at Chalmers University in 2006. His research has been in multiphase flow.

Berend van Wachem is a Reader at Imperial College London in the UK. His research projects involve multiphase flow modelling, ranging from understanding the behaviour of turbulence on the scale of individual particles to the large-scale modelling of gas–solid and gas–liquid flows.

Computational Fluid Dynamics for Engineers

BENGT ANDERSSON
Chalmers University, Sweden

RONNIE ANDERSSON
Chalmers University, Sweden

LOVE HÅKANSSON
Engineering Data Resources – EDR, Norway

MIKAEL MORTENSEN
Norwegian Defence Research Establishment, Norway

RAHMAN SUDIYO
University of Gadjah Mada, Indonesia

BEREND VAN WACHEM
Imperial College London, UK

CAMBRIDGE
UNIVERSITY PRESS

University Printing House, Cambridge CB2 8BS, United Kingdom

Published in the United States of America by Cambridge University Press, New York

Cambridge University Press is part of the University of Cambridge.

It furthers the University's mission by disseminating knowledge in the pursuit of education, learning and research at the highest international levels of excellence.

www.cambridge.org
Information on this title: www.cambridge.org/9781107018952

First published 2012

A catalogue record for this publication is available from the British Library

Library of Congress Cataloguing in Publication data
Computational fluid dynamics for engineers / Bengt Andersson . . . [et al.].
p. cm.
Includes bibliographical references and index.
ISBN 978-1-107-01895-2 (hardback)
1. Fluid dynamics. 2. Engineering mathematics. I. Andersson, Bengt, 1947 June 15–
TA357.C58776 2011
532.05 – dc23 2011037992

ISBN 978-1-107-01895-2 Hardback

Additional resources for this publication at www.cambridge.org/9781107018952

Contents

Preface

Computational fluid dynamics (CFD) has become an indispensable tool for engineers. CFD simulations provide insight into the details of how products and processes work, and allow new products to be evaluated in the computer, even before prototypes have been built. It is also successfully used for problem shooting and optimization. The turnover time for a CFD analysis is continuously being reduced since computers are becoming ever more powerful and software uses ever more efficient algorithms. Low cost, satisfactory accuracy and short lead times allow CFD to compete with building physical prototypes, i.e. 'virtual prototyping'.

There are many commercial programs available, which have become easy to use, and with many default settings, so that even an inexperienced user can obtain reliable results for simple problems. However, most applications require a deeper understanding of fluid dynamics, numerics and modelling. Since no models are universal, CFD engineers have to determine which models are most appropriate to the particular case. Furthermore, this deeper knowledge is required since it gives the skilled engineer the capability to judge the potential lack of accuracy in a CFD analysis. This is important since the analysis results are often used to make decisions about what prototypes and processes to build.

Our ambition is that this book will provide sufficient background for CFD engineers to solve more advanced problems involving advanced turbulence modelling, mixing, reaction/combustion and multiphase flows. This book presents the equations that are to be solved, *but, more importantly*, the essential physics in the models is described, and the limitations of the models are discussed. In our experience, the most difficult part for a CFD engineer is not to select the best numerical schemes but to understand the fluid dynamics and select the appropriate models. This approach makes the book useful as an introduction to CFD irrespective of the CFD code that is used, e.g. finite-volume, finite-element, lattice Boltzmann etc.

This book requires a prior knowledge of transport phenomena and some understanding of computer programming. The book (and the tutorials/project) is primarily intended for engineering students. The objective is to teach the students how to do CFD analysis correctly but not to write their own CFD code. Beyond this, the book will give an understanding of the strengths and weakness of CFD simulations. The book is also useful for experienced and practicing engineers who want to start using CFD themselves or, as project managers, purchase these services from consulting firms.

We have added several questions of reflective character throughout the book; it is recommended that you read these to confirm that the most important parts have been understood. However, the book intentionally contains few simulation results and worked-through examples. Instead we have developed three tutorials and one larger project that give students the required hands-on experience. The tutorials take 6–8 hours each to run; the project takes 30 hours to run and write a report. These tutorials and the project are available from the authors. We have chosen to use a commercial code (ANSYS/Fluent) in our course, but the problem formulations are written very generally and any commercial program could be used. Our experience is that commercial CFD programs can be obtained for teaching purposes at very low cost or even free. An alternative is to use an open-source program, e.g. OpenFoam. Unfortunately, the user interface in OpenFoam is not as well developed as are those in the commercial programs, so students will have additional problems in getting their programs working.

This book has successfully been used in a CFD course at Chalmers University since 2004. Every year approximately 60 chemical and mechanical engineering students take the course. Over the years this book has also been used for PhD courses and courses in the industry. The text has been rewritten every year to correct errors and in response to very valuable suggestions from the students. PowerPoint lecture notes are also available from the authors.

Scope

Chapter 1 provides an introduction to what can be solved with a CFD program and what inputs are required from the user. It also gives an insight into what kind of problems are easy or difficult to solve and how to obtain reliable results.

Chapter 2 contains the equations that are solved by the CFD software. The student should know these equations from their prior courses in transport phenomena, but we have included them because they are the basis for CFD and an up-to-date knowledge of them is essential.

In *Chapter 3* the most common numerical methods are presented and the importance of boundedness, stability, accuracy and convergence is discussed. We focus on the finite volumes on which most commercial software is based and only a short comparison with the finite-element method is included. There is no best method available for all simulations since the balance among stability, accuracy and speed depends on the specific task.

Chapter 4 gives a solid introduction to turbulence and turbulence modelling. Since simulation of turbulent flows is the most common application for engineers, we have set aside a large part of the chapter to describe the physics of turbulence. With this background it is easier to present turbulence modelling, e.g. why sources and sinks for turbulence are important. The k–ε model family, k–ω, Reynolds stress and large-eddy models are presented and boundary conditions are discussed in detailed.

Chapter 5 carefully analyses turbulent mixing, reaction and combustion. The physics of mixing is presented and the consequence of large fluctuations in concentration is discussed. A probability distribution method is presented and methods to solve instantaneous, fast and slow kinetics are formulated. A simple eddy-dissipation model is also presented.

In *Chapter 6* multiphase models are presented. First various tools with which to characterize multiphase flow and forces acting on particles are presented. Eulerian–Lagrangian, Euler–Euler, mixture (algebraic slip), volume-of-fluid and porous-bed models are presented and various closures for drag, viscosity etc. are formulated. Simple models for mass and heat transfer between the phases are also presented.

Finally, *Chapter 7* contains a best-practice guideline. It is based on the guidelines presented by the European Research Community on Flow Turbulence and Combustion, ERCOFTAC, in 2000 and 2009 for single-phase and multiphase systems, respectively.

In *Tutorial 1* reactions inside a spherical porous catalyst particle are studied. The reaction is exothermic and flow, heat and species must be modelled. The student will learn how to draw and mesh a two-dimensional (2D) geometry. They will also specify boundary conditions and select the models to solve. The kinetics is written as a user-defined function (UDF) and the student will learn how to implement a UDF in a CFD simulation. Convergence is a problem, and the student will learn about physical reaction instability, numerical instabilities, under-relaxation and numerical diffusion. In the report the student is required to show that their simulations fulfil the criteria given in the best practice guidelines.

In *Tutorial 2* turbulent mixing and combustion in a bluff-body stabilized non-premixed turbulent flame is studied. An instantaneous adiabatic equilibrium reaction, i.e. combustion of methane in air, is simulated. The student should select an appropriate turbulence model and solve for flow, turbulence, mean mixture fraction, mixture fraction variance, species and heat with appropriate boundary conditions, e.g. wall functions. Mesh adaptation to obtain the proper y^+ for the wall functions is introduced. The students should analyse whether jets and recirculations exist in the flow and whether the reaction is fast compared with turbulent mixing.

In *Tutorial 3* a spray is modelled using an Eulerian–Lagrangian multiphase model. Continuous phase and spray velocity combined with heat transfer and evaporation are modelled. The student should analyse the fluid–spray interaction and choose what forces should be included in the model.

The *Project* is dedicated to the design of an industrial-scale selective catalytic reduction (SCR) process. The student generates a three-dimensional (3D) computer-aided design (CAD) model and mesh, analyses the performance, and suggests and evaluates design improvements of the SCR reactor.

Acknowledgements

We are very grateful to the students who have given us very valuable feedback and thus helped us to improve the book. We would also like to thank Mrs Linda Hellström, who did the graphics, and Mr Justin Kamp, who corrected most of our mistakes in the English language.

Bengt Andersson

1 Introduction

The purpose of this chapter is to explain the input needed to solve CFD problems, e.g. CAD geometry, computational mesh, material properties, boundary conditions etc. The difficulty and accuracy of CFD simulations for various applications, such as laminar and turbulent flows, single-phase and multiphase flows and reactive systems are discussed briefly.

1.1 Modelling in engineering

Traditional modelling in engineering is heavily based on empirical or semi-empirical models. These models often work very well for well-known unit operations, but are not reliable for new process conditions. The development of new equipment and processes is dependent on the experience of experts, and scaling up from laboratory to full scale is very time-consuming and difficult. New design equations and new parameters in existing models must be determined when changing the equipment or the process conditions outside the validated experimental database. A new trend is that engineers are increasingly using computational fluid dynamics (CFD) to analyse flow and performance in the design of new equipment and processes. CFD allows a detailed analysis of the flow combined with mass and heat transfer. Modern CFD tools can also simulate transport of chemical species, chemical reactions, combustion, evaporation, condensation and crystallization.

1.2 CFD simulations

Simple, single-phase laminar flow can be simulated very accurately, and for most single-phase turbulent flows the simulations are reliable. However, many systems in engineering are very complex, and simulations of multiphase systems and systems including very fast reactions are at present not very accurate. For these systems, the traditional models using well-proven design equations that have been verified over many years are more accurate than the best CFD simulations. However, design equations are available only for existing equipment and for a limited range of process conditions, and CFD simulations can be very useful even when no accurate predictions are possible, e.g. by calibrating the model to get a solution that is verified experimentally. From this simulation we can

Table 1.1 Potential CFD simulations in engineering

Flow	Mass transfer	Heat transfer
Laminar	Convection	Convection
Turbulent	Diffusion	Conduction
Single-phase	Reaction	Radiation
Multiphase	Phase transfer	

do parameter studies by implementing small changes in the parameters, e.g. to assess what will result from changes in temperature, flow, viscosity etc.

One advantage using CFD is that it is possible to obtain detailed local information on the simulated system. In a fluidized bed it is possible to simulate not only the conversion but also the local temperature, the entrainment of particles, the backmixing and bubble formation. This detailed information will help by building a qualitative understanding of the process, and a parameter study can reveal additional information such as the bottle necks and the operational limits of the equipment.

CFD simulation without proper knowledge can be a very uncertain tool. The commercial CFD programs have many default settings and will almost always give results from the simulations, but, to obtain reliable results, the model must be chosen with a logical methodology. A converged solution displays the results of the specific chosen model with the given mesh; however, it does not reveal the truth. Without proper understanding of the CFD program and the modelling theory behind it, CFD can become limited to 'colourful fluid display'.

1.3 Applications in engineering

Virtual prototyping is now the standard method to develop new products in e.g. the automotive industry. This very effective method is now being introduced into other fields of engineering. Within chemical engineering we find applications in most fields, e.g. reactor modelling, separations and heat transfer. Unfortunately, we seldom have single-phase laminar flow that can easily be simulated. Chemical reactors are almost always turbulent and often multiphase. Mixing of the reactants and removal of the heat produced are the main objectives for typical reactors such as stirred-tank reactors, bubble columns, trickle beds and fluidized beds. Almost all separations are multiphase, e.g. distillations, extractions, filtering and crystallization. Most heat-transfer equipment involves single-phase flow, yet boiling and condensation are also commonly used. Modern CFD programs can simulate a very wide range of systems and Table 1.1 lists possible simulations in CFD.

1.4 Flow

It is useful to separate the properties of fluids and flows. The properties of fluids, e.g. viscosity, density, surface tension, diffusivity and heat conduction, are intrinsic properties that can be described as functions of temperature, pressure and composition. Properties that depend on the flow include pressure, turbulence and turbulent viscosity.

From a CFD modelling point of view it is useful to separate possible flows into the following categories:

laminar–turbulent
steady–transient
single-phase–multiphase

1.4.1 Laminar flow

In laminar flow the Navier–Stokes equations describe the momentum transport of flow that is dominated by viscous forces. It is possible with CFD to obtain very accurate flow simulations for single-phase systems, provided that the flow is always laminar. The transitions between laminar and turbulent flow, both from turbulent to laminar and from laminar to turbulent, are difficult to simulate accurately. In this region the flow can fluctuate between laminar and turbulent and turbulent slugs can frequently appear in laminar flow far below the critical Reynolds number for transition to turbulent flow.

Simulation of heat transfer is also often very accurate and a good prediction of temperature can easily be obtained. Mass transfer in the gas phase is also quite straightforward. However, the diffusivities in liquids are about four orders of magnitude lower than those in the gas phase at atmospheric pressure and accurate mass-transport simulations in laminar liquids are difficult. An estimation of the transport distance due to diffusion in laminar flow can be calculated from $x = \sqrt{Dt}$. The diffusivity in liquids is of the order of 10^{-9} m^2 s^{-1} and the average transport distance in 1 s is about 3 μm, i.e. very dense grids are required. The corresponding transport distance in the gas phase is 300 μm, with a gas-phase diffusivity of 10^{-5} m^2 s^{-1}.

1.4.2 Turbulent flow

The Navier–Stokes equations describe turbulent flows, but, due to the properties of the flow, it is seldom possible to solve the equations for real engineering applications even with supercomputers. In a stirred-tank reactor the lifetime and size of the smallest turbulent eddies, the Kolmogorov scales, are about 5 ms and 50 μm, respectively. A very fine time and space resolution is needed when solving the Navier–Stokes equations and it is at present not possible. Direct solution of the Navier–Stokes equations (DNS) for small systems is, however, very useful for developing an understanding of turbulence, and for developing new models.

A more cost-effective method is to resolve only the large-scale turbulence, by filtering out the fine-scale turbulence, and model these small scales as flow-dependent effective viscosity. This method, large-eddy simulation (LES), is growing in popularity since it makes it possible to simulate simple engineering flows on a fast PC. The simulations are, however, very time consuming on a PC, and several weeks can often be needed to obtain good statistical averages even for rather simple flows.

For more complex flows, it is not possible to resolve the turbulence fluctuations at all. Most engineering simulations are done with Reynolds-averaged Navier–Stokes (RANS) methods. In these models the turbulent fluctuations are time averaged, yet reasonable velocity averages can be simulated from these models. However, there are important properties of the flow that are not resolved. Everything that occurs on a scale below the size of the grid is not resolved, e.g. mixing of chemical species or the break-up and coalescence of bubbles and drops in multiphase flow. Additional models must be added to the RANS models in order to include these phenomena.

1.4.3 Single-phase flow

In single-phase laminar flow we can obtain very accurate solutions and in turbulent flow we can in most cases obtain satisfactory flow simulations. The main problem is usually simulation of the mixing of reactants for fast reactions in laminar or turbulent flow. When the reaction rate is fast compared with mixing, there will be strong concentration gradients that cannot be resolved in the grid, and a model for mixing coupled with a chemical reaction must be introduced. Combustion in the gas phase and ion–ion reactions in the liquid phase belong to this category.

1.4.4 Multiphase flow

Multiphase flow may consist of gas–liquid, gas–solid, liquid–liquid, liquid–solid or gas–liquid–solid systems. For a multiphase system containing very small particles, bubbles or drops that follow the continuous phase closely, reasonable simulation results can be obtained. Systems in which the dispersed phase has a large effect on the continuous phase are more difficult to simulate accurately, and only crude models are available for multiphase systems with a high load of the dispersed phase. At the moment, the quality of the simulations is limited not by the computer speed or memory but by the lack of good models for multiphase flow. However, multiphase flows are very important in engineering since many common processes involve multiphase flow.

The mass and heat transfer between the phases are of interest in many applications, e.g. in boiling, heterogeneous catalysis and distillation. For these simulations we must introduce empirical or semi-empirical correlations to describe mass and heat transfer. The mass- and heat-transfer coefficients are usually calculated from the traditional correlations for the Sherwood, *Sh*, and Nusselt, *Nu*, numbers. The advantage with CFD is that *Sh* and *Nu* can be computed using local flow properties. The mass and heat transfer are also affected by the coalescence and break-up of bubbles and drops. The phenomenon of break-up and coalescence is not included in this book since only very simple models are available for simulation of the effect of turbulence and shear rate on drop or bubble size distributions.

1.5 CFD programs

There are many commercial general-purpose CFD programs available, e.g. Fluent, CFX, Star-CD, FLOW-3D and Phoenics. There are also some very specialized programs

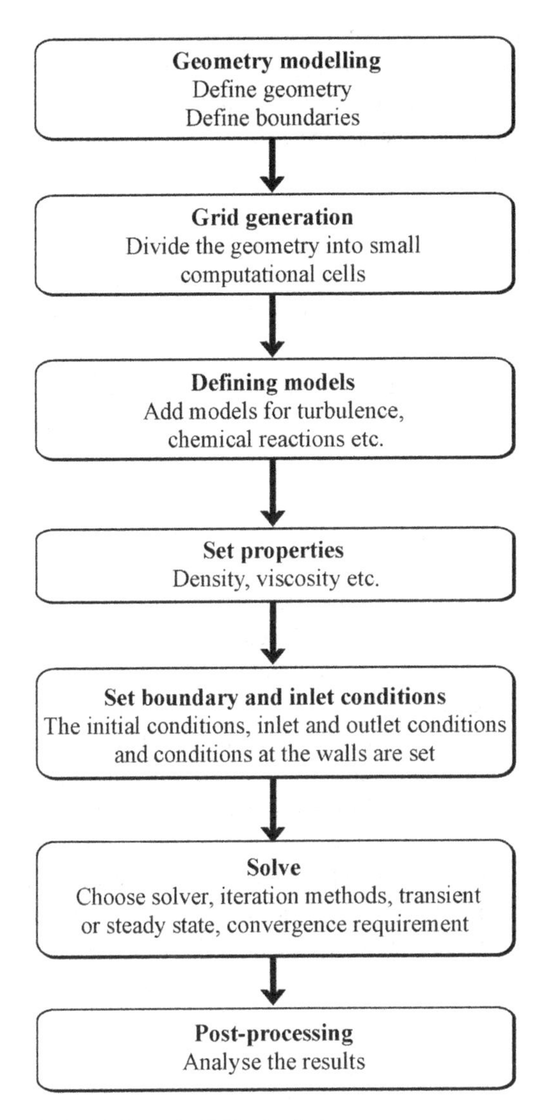

Figure 1.1 Steps in CFD simulations.

simulating combustion in engines, cooling of semiconductors and simulation of weather. A very useful open-source program that can handle most CFD problems is OpenFoam. However, the documentation and the user interface are not as well developed as those for the commercial codes.

Commercial CFD packages contain modules for CAD drawing, meshing, flow simulations and post-processing. In solving a problem using CFD there are many steps that must be defined, as is illustrated in Figure 1.1.

Geometry modelling

Solving a CFD problem starts with a two-dimensional (2D) or three-dimensional (3D) drawing of the geometry of the system. A CAD program is included in all commercial

CFD programs but the geometry of the system can usually be drawn in any CAD program and imported into the grid-generation program. However, CAD programs not designed for CFD often contains details that cannot be included in CFD drawings, e.g. nuts and bolts. These drawings must be cleaned before they can be used in a meshing program.

Grid generation (meshing)

The equations for momentum transport are nonlinear, which means that the computational volume must be discretized properly to obtain an accurate numerical solution of the equations. Accurate meshing of the computational domain is as important as defining the physical models. An ill-conditioned mesh can give rise to very inaccurate results, so the quality of the mesh, e.g. its aspect ratio and skewness, must be evaluated prior to the simulations. Most CFD programs have also the possibility of adaptation, which, after a preliminary result has been obtained, enables local refinement of the grid where necessary.

Define models

For single-phase laminar flow, the Navier–Stokes equations can be solved directly, but for turbulent and multiphase flows the user must select the most appropriate model. There are few generally accepted models for turbulence and multiphase flow, but there are hundreds of models to choose from. For each model there are also several parameters that must be set. Usually the default values are the best choice, but in some cases the user can find more suitable parameters. In most commercial CFD programs it is also possible to write your own model as a user-defined subroutine/function (UDS/UDF) in Fortran or C. Not all properties are resolved in the CFD program, and many semi-empirical models must be defined. The momentum, heat and mass transfer between the dispersed and continuous phases in multiphase flow are defined by empirical models for drag and Sherwood and Nusselt numbers as functions of the local particle Reynolds number or turbulence intensity.

Set properties

All physical properties of the fluids must be defined, e.g. the viscosity and density and their temperature, composition and pressure dependence. Some are built into the CFD software or easily found in available databases. The programs also provide polynomials for which you can add your own constants. It is also possible to write a UDS/UDF that is added to the CFD program for calculating the properties.

Set boundary and initial conditions

All inlet and outlet conditions must be defined, as must conditions on the walls and other boundaries. Rotational symmetry and other symmetries, e.g. periodic induced boundary conditions, must also be defined. Periodic boundary conditions are very useful with rotating equipment, e.g. a turbine, when only a part is modelled. Initial conditions for transient simulations or an initial guess to start the iterations for steady-state simulations must also be provided.

Solve

For the solver you can choose either a segregated or a coupled solver, pressure- or density-based, and for unsteady problems you must choose either implicit or explicit time-stepping methods. Numerical schemes to enhance convergence, e.g. multigrid, upwind schemes or under-relaxation factors, must be defined. The quality of an acceptable solution in terms of the convergence criteria must also be defined.

Post-processing/analysis

The first objective in the post-processing is to analyse the quality of the solution. Is the solution independent of the grid size, the convergence criterion and the numerical schemes? Have the proper turbulence model and boundary conditions been chosen, and is the solution strongly dependent on those choices?

Analysis of the final simulation results will then give local information about flow, concentrations, temperatures, reaction rates etc.

For very complex systems the results are not very accurate, but CFD can still be very useful in answering the question 'What if?'. Starting with an adjusted CFD simulation that fits the available experimental results, a parameter study can reveal how e.g. the mixing is affected by an increase in molecular viscosity.

Questions

(1) What can be simulated in a CFD program?
(2) Why is it not possible to solve Navier–Stokes equations for turbulent flows for typical engineering applications?
(3) What steps are involved in solving a typical CFD problem?

2 Modelling

In CFD the equations for continuity, momentum, energy and species are solved. These coupled partial nonlinear differential equations are in general not easy to solve numerically and analytical solutions are available for only very few limited cases. The reader is expected to have a basic knowledge of transport phenomena but, since all CFD is based on these few equations, they are given here in tensor notation so that the reader can become familiar with this notation.

A general balance formulation in tensor notation for a scalar, vector or tensor ϕ can be formulated as

$$\frac{\partial \phi}{\partial t} + U_i \frac{\partial \phi}{\partial x_i} = D \frac{\partial^2 \phi}{\partial x_i \partial x_i} + S(\phi), \tag{2.1}$$

where the terms have the following meanings:

$$\left\{ \begin{array}{l} \text{rate of} \\ \text{accumulation} \end{array} \right\} + \left\{ \begin{array}{l} \text{transport by} \\ \text{convection} \end{array} \right\} = \left\{ \begin{array}{l} \text{transport by} \\ \text{diffusion} \end{array} \right\} + \left\{ \begin{array}{l} \text{source} \\ \text{terms} \end{array} \right\}$$

This notation will be used throughout the book and the reader must be familiar with this notation. In this convention there is an understood summation that is written explicitly below (see the appendix for further information):

$$\frac{\partial \phi}{\partial t} + \sum_i U_i \frac{\partial \phi}{\partial x_i} = \sum_i D \frac{\partial^2 \phi}{\partial x_i^2} + S(\phi)$$

In 3D Cartesian coordinates i can take the values 1, 2 and 3, and for a scalar ϕ the equation above becomes

$$\frac{\partial \phi}{\partial t} + U_1 \frac{\partial \phi}{\partial x_1} + U_2 \frac{\partial \phi}{\partial x_2} + U_3 \frac{\partial \phi}{\partial x_3} = D \left(\frac{\partial^2 \phi}{\partial x_1^2} + \frac{\partial^2 \phi}{\partial x_2^2} + \frac{\partial^2 \phi}{\partial x_3^2} \right) + S(\phi). \tag{2.2}$$

Note that since ϕ is a scalar there is only one equation describing how ϕ is distributed in the three dimensions x_1, x_2 and x_3. The easiest way to understand the notation is to identify whether the dependent variable is a scalar, vector or tensor. When ϕ is a scalar, e.g. temperature, T, only one equation is possible, since there is only one temperature at a given position. When ϕ is a vector, e.g. $\phi = [U_1 U_2 U_3]^{\mathrm{T}}$, there will be one equation for each of the three velocities. Equation (2.21) below is a tensor notation of the three momentum equations written in Eq. (2.20). For a tensor τ_{ij} there will be nine equations in three dimensions, cf. Eq. (2.7) below.

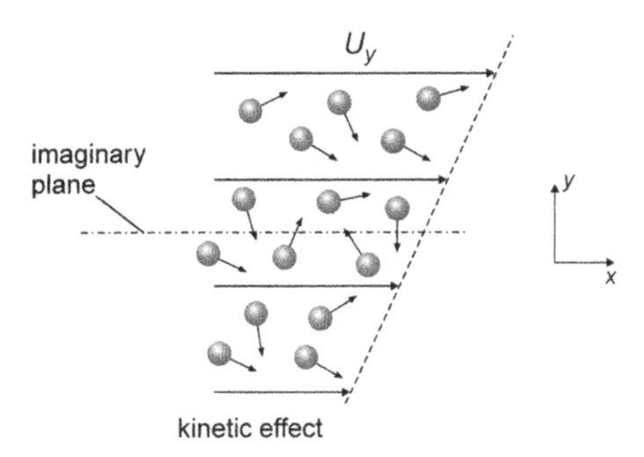

Figure 2.1 The effect of random movement of molecules on momentum transfer.

Accumulation, convection, diffusion and a source term will appear in many equations, and by identifying these terms it will be easier to refer to the various parts in the equations. The accumulation is recognized from the time derivative, the convection term from the velocity term and the first-order derivative, and the diffusion term from the second-order derivative and the transport coefficient, e.g. diffusivity, conductivity, or viscosity. The source term is a function solely of the local variables.

2.1 Mass, heat and momentum balances

All modelling is easier when the underlying physics can be understood and momentum balances are the basis of all fluid dynamics. Transport of mass, heat and momentum occurs by convection of the mean flow and by random movement of molecules or in turbulent flow by random movement of fluid elements. Viscous transport of momentum is due to the random movement of molecules carrying their average momentum in all directions as shown in Figure 2.1.

2.1.1 Viscosity, diffusion and heat conduction

There are many similarities among viscosity, diffusion and heat conduction in fluids. The mechanism for transport in all these cases is random movement of molecules or fluid elements. According to the kinetic theory of gases a molecule moves randomly in all directions, giving a mean velocity of $\overline{u} = \sqrt{8RT/(\pi M_v)}$. This molecule will move a distance corresponding to the mean free path $\lambda = k_B T/(\sqrt{2\pi d^2 P})$ before it collides with another molecule and transfers momentum and heat to that molecule. For oxygen at room temperature and atmospheric pressure, $\overline{u} = 444$ m s^{-1} and $\lambda = 71.4$ nm. For an ideal gas the kinetic viscosity, the diffusivity and the heat diffusivity are all of the same order, $\nu \approx D \approx D_H \approx \frac{1}{3}u\lambda$. Heat conduction is related to heat diffusivity by the amount of energy that each molecule carries, $k = \rho c_p D_H$. From this simple model one can also observe that, for gases, the viscosity depends on the temperature, pressure, relative molecular mass, M_v, and size of the molecule, d. The Schmidt number $Sc = \nu/D$

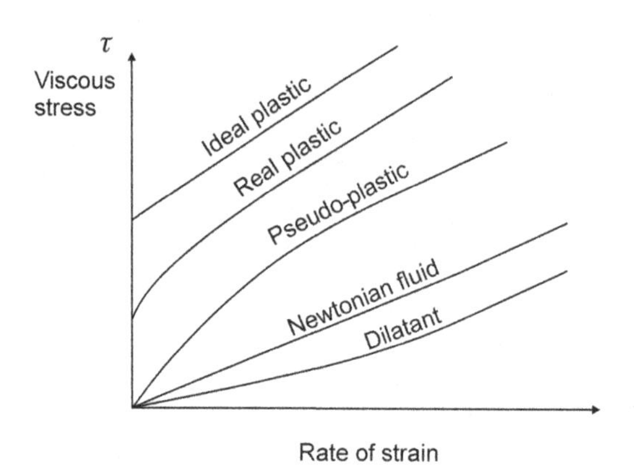

Figure 2.2 The viscous stress for Newtonian and non-Newtonian fluids.

and the Prandtl number $Pr = \rho c_p \nu / k$ describe the ratio between viscosity and diffusivity and that between viscosity and heat conduction, respectively. Both Sc and Pr are of the order of 0.7 for air.

In liquids, the transport is somewhat different. The molecules are in close contact with neighbouring molecules and the movement of the molecules can be modelled as movement with viscous drag. Momentum and heat are transported much faster than mass, because the momentum and heat can be transferred to other molecules by collisions, while diffusion is limited to the movement of the single molecules. This difference can be seen by noting that $Sc \approx 1000$ while $Pr \approx 7$ for transport in water.

The mechanism for momentum, mass and heat transport is similar in turbulent flow. Here the random movement of turbulent eddies will transfer fluid elements containing momentum, species and energy. The turbulent viscosity is of the order of the turbulent velocity times the average distance travelled by a turbulent eddy. Since all transport is by the turbulent eddies, the turbulent Sc and Pr numbers are all of the order of unity, both for gases and for liquids. The kinetic theory of gases is also the governing idea in some of the models for viscosity in multiphase flow, e.g. the kinetic theory for granular flow (KTGF).

Newton's law of viscosity

The viscous stress as a function of velocity gradients can vary significantly for different fluids depending on how the molecules arrange themselves when exposed to strain, as shown in Figure 2.2.

The simplest fluid is the Newtonian fluid, and fortunately many of the common fluids are very close to Newtonian, e.g. gases, water and other simple liquids. In Newtonian fluids the viscous stress is a linear function of the rate of strain. In Figure 2.3 a simple laminar flow with velocity U_1 flows in only one dimension x_1. The resistance of the flow that is observed in a pressure drop is due to the fact that the momentum in the velocity direction x_1 is transported in the x_2 direction due to viscous forces.

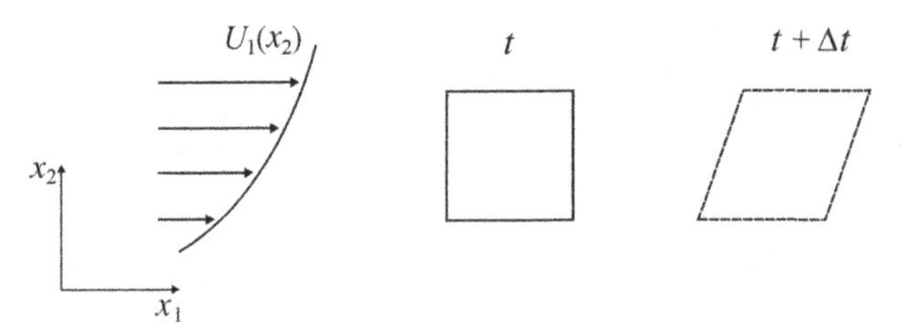

Figure 2.3 The distortion of a fluid element due to strain rate dU_1/dx_2.

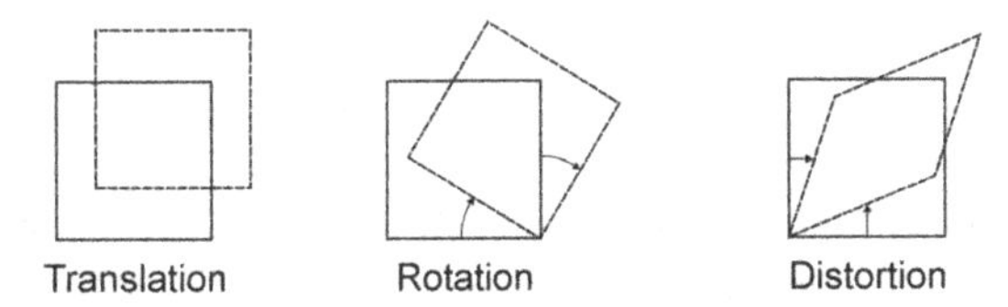

Figure 2.4 Translation, rotation and distortion of a fluid element.

For a Newtonian fluid the linear dependence between stress and the velocity gradient is expressed as

$$\tau_{21} = \mu \frac{\mathrm{d}U_1}{\mathrm{d}x_2}. \tag{2.3}$$

Here the first index of τ_{21} denotes the direction of transport and the second the direction of the momentum. Note that the stress tensor is written with a positive sign. There is a tradition in chemical engineering of viewing the stress tensor as the transport of momentum and consequently defining it with a negative sign since the direction of momentum transport is opposite to the direction of the gradient analogous to heat and mass transfer.

The velocity gradient in itself does not cause the stresses, but rather the stresses arise due to the distortion of the fluid element. A pure translation or rotation of the element will not give rise to viscous stress (Figure 2.4).

In a 2D case, the viscous stress will depend on distortion of the fluid element and the viscous stress becomes a linear function of the strain rate

$$\tau_{12} = \tau_{21} = \mu \left(\frac{\partial U_1}{\partial x_2} + \frac{\partial U_2}{\partial x_1} \right). \tag{2.4}$$

The expression is symmetric in the two dimensions and the two stresses must be equal, $\tau_{12} = \tau_{21}$.

The normal stresses are also affected by the compression of the fluid elements,

$$\tau_{11} = 2\mu \frac{\partial U_1}{\partial x_1} - \left(\tfrac{2}{3}\mu - \kappa\right) \left(\frac{\partial U_1}{\partial x_1} + \frac{\partial U_2}{\partial x_2} \right) \tag{2.5}$$

and

$$\tau_{22} = 2\mu \frac{\partial U_2}{\partial x_2} - \left(\tfrac{2}{3}\mu - \kappa\right) \left(\frac{\partial U_1}{\partial x_1} + \frac{\partial U_2}{\partial x_2} \right). \tag{2.6}$$

In 3D the nine possible stresses become

$$\tau_{ij} = \tau_{ji} = \mu \left(\frac{\partial U_i}{\partial x_j} + \frac{\partial U_j}{\partial x_i} \right) - \left(\tfrac{2}{3}\mu - \kappa \right) \delta_{ij} \left(\frac{\partial U_k}{\partial x_k} \right), \tag{2.7}$$

where δ_{ij} is the Kronecker delta

$$\delta_{ij} = (\mathbf{I})_{ij} = \begin{cases} 1 & \text{if } i = j, \\ 0 & \text{if } i \neq j. \end{cases}$$

Sometimes the pressure is added to the normal stresses and the stress tensor is written

$$\sigma_{ij} = \mu \left(\frac{\partial U_1}{\partial x_1} + \frac{\partial U_2}{\partial x_2} \right) - \left(\tfrac{2}{3}\mu - \kappa \right) \delta_{ij} \left(\frac{\partial U_k}{\partial x_k} \right) - \delta_{ij} P. \tag{2.8}$$

The dilatational viscosity κ is important only for shock waves and sound waves. From the kinetic theory of gases it has also been shown that κ is zero for monatomic gases at low pressure. In this book we will mostly describe non-compressible fluids, for which the term $\partial U_k / \partial x_k$ on the right-hand side is zero and the viscous stress for a Newtonian fluid is described by

$$\tau_{ij} = \tau_{ji} = \mu \left(\frac{\partial U_i}{\partial x_j} + \frac{\partial U_j}{\partial x_i} \right). \tag{2.9}$$

2.2 The equation of continuity

A material balance over a stationary fluid element $\Delta x_1 \Delta x_2 \Delta x_3$ is written

$$\{\text{accumulation}\} = \{\text{transport in}\} - \{\text{transport out}\},$$

$$\begin{aligned} \Delta x_1\, \Delta x_2\, \Delta x_3 \frac{\partial \rho}{dt} = {} & \Delta x_2\, \Delta x_3 \left[(\rho U_1)|_{x_1} - (\rho U_1)|_{x_1 + \Delta x_1} \right] \\ & + \Delta x_1\, \Delta x_3 \left[(\rho U_2)|_{x_2} - (\rho U_2)|_{x_2 + \Delta x_2} \right] \\ & + \Delta x_1\, \Delta x_2 \left[(\rho U_3)|_{x_3} - (\rho U_3)|_{x_3 + \Delta x_3} \right]. \end{aligned} \tag{2.10}$$

See Figure 2.5. This formulation corresponds to the finite-volume formulation of the continuity equation for a hexahedral mesh aligned with the coordinate axis. Dividing by $\Delta x_1\ \Delta x_2\ \Delta x_3$ and taking the limit $\Delta x \to 0$ gives the continuity equation

$$\frac{\partial \rho}{\partial t} + \frac{\partial \rho U_1}{\partial x_1} + \frac{\partial \rho U_2}{\partial x_2} + \frac{\partial \rho U_3}{\partial x_3} = 0 \quad \text{or} \quad \frac{\partial \rho}{\partial t} = -(\nabla \cdot \rho \mathbf{U}). \tag{2.11}$$

Tensor notation is mostly applied to equations in the book and the equation above is written

$$\frac{\partial \rho}{\partial t} + \frac{\partial \rho U_j}{\partial x_j} = 0. \tag{2.12}$$

In 3D the index j is 1, 2 and 3, giving Eq. (2.11).

The continuity equation is difficult to solve numerically. In CFD programs, the continuity equation is often combined with the momentum equations (see below) to form

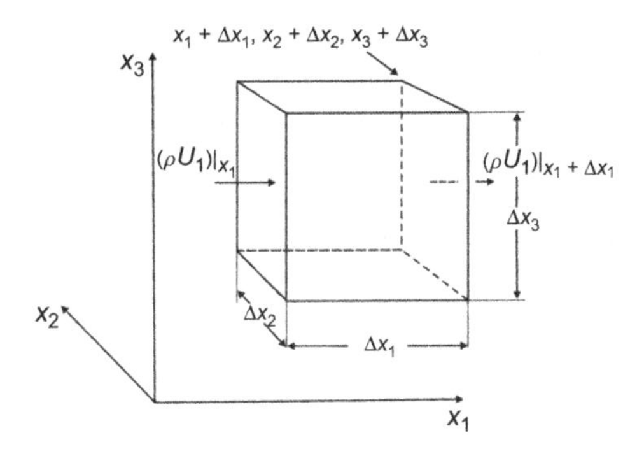

Figure 2.5 A material balance over a fluid element.

a Poisson equation for pressure. For constant density and viscosity the new equation will be

$$\frac{\partial}{\partial x_i}\left(\frac{\partial P}{\partial x_i}\right) = -\frac{\partial}{\partial x_i}\left[\frac{\partial(\rho U_i U_j)}{\partial x_j}\right]. \tag{2.13}$$

This equation has more suitable numerical properties and can be solved by proper iteration methods. The Navier–Stokes equations are used for simulation of velocity and pressure. The Poisson formulation introduces pressure as a dependent variable, and the momentum equations can be formulated to solve for velocity. The Poisson equation and other methods to solve the continuity are discussed in Chapter 3.

In many cases it is more convenient to describe the change in flow of a fluid element that moves with the flow. On performing the derivation in Eq. (2.12) we obtain

$$\frac{\partial \rho}{\partial t} + U_1\frac{\partial \rho}{\partial x_1} + U_2\frac{\partial \rho}{\partial x_2} + U_3\frac{\partial \rho}{\partial x_3} = -\rho\left(\frac{\partial U_1}{\partial x_1} + \frac{\partial U_2}{\partial x_2} + \frac{\partial U_3}{\partial x_3}\right). \tag{2.14}$$

The left-hand side is the substantial derivative of density, i.e. the time derivative for a fluid element that follows the fluid motion. The equation can be abbreviated as

$$\frac{\mathrm{D}\rho}{\mathrm{D}t} = -\rho\frac{\partial U_i}{\partial x_i} \tag{2.15}$$

and the substantial operator defined as

$$\frac{\mathrm{D}}{\mathrm{D}t} \equiv \frac{\partial}{\partial t} + U_1\frac{\partial}{\partial x_1} + U_2\frac{\partial}{\partial x_2} + U_3\frac{\partial}{\partial x_3}. \tag{2.16}$$

Incompressible flow is defined by having constant density along the streamline, i.e. the left-hand side of Eq. (2.15) is zero and

$$\frac{\partial U_i}{\partial x_i} = 0. \tag{2.17}$$

The assumption of incompressible flow will simplify the modelling substantially, and we will use it throughout the book. Truly incompressible flow does not exist, but the assumption of incompressible flow is valid for most engineering applications. A local change in pressure will spread with the speed of sound, and, when modelling phenomena

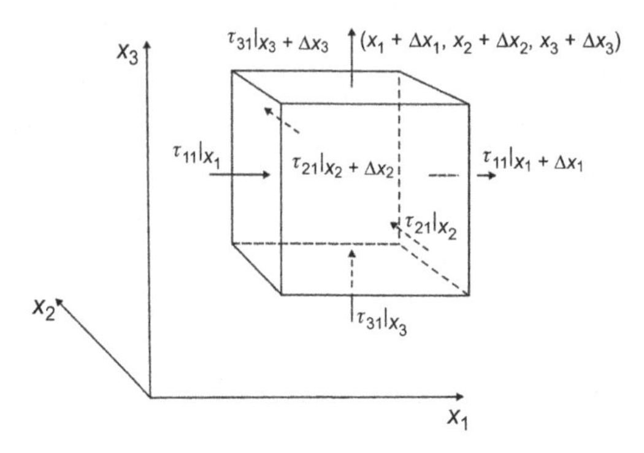

Figure 2.6 Momentum balance over a fluid element.

much slower than the speed of sound, we can safely assume that the new pressure is reached in each time step, i.e. compressible flows can be characterized by the value of the Mach number $M = U/c$. Here, c is the speed of sound in the gas $c = \sqrt{\gamma RT}$, where γ is the ratio of specific heats c_p/c_v. At Mach numbers much less than 1.0 ($M < 0.1$), compressibility effects are negligible and the variation of the gas density due to pressure waves can safely be ignored in the flow modelling. The density change due to pressure drops and temperature variations will automatically be compensated for by the state equation that describes how the density is related to pressure and temperature.

2.3 The equation of motion

The momentum balance for the U_1 momentum over the volume $\Delta x_1 \Delta x_2 \Delta x_3$ is given by

$$\left\{\begin{array}{l}\text{rate of}\\ \text{momentum}\\ \text{accumulation}\end{array}\right\} = \left\{\begin{array}{l}\text{rate of}\\ \text{momentum}\\ \text{in}\end{array}\right\} - \left\{\begin{array}{l}\text{rate of}\\ \text{momentum}\\ \text{out}\end{array}\right\} + \left\{\begin{array}{l}\text{sum of forces}\\ \text{acting on}\\ \text{the system}\end{array}\right\}. \tag{2.18}$$

Newton's law requires that the change in momentum in each direction should be balanced by the forces acting in that direction. The arrows in Figure 2.6 describe the direction of viscous transport of U_i momentum. A balance for the momentum of the velocity component in the x_1 direction, i.e. U_1, is written

$$\begin{aligned}\Delta x_1\,\Delta x_2\,\Delta x_3 \frac{\partial \rho U_1}{\partial x_1} &= \Delta x_2\,\Delta x_3\left[(\rho U_1 U_1)|_{x_1} - (\rho U_1 U_1)|_{x_1+\Delta x_1}\right] - \tau_{11}|_{x_1} + \tau_{11}|_{x_1+\Delta x_1}\\ &+ \Delta x_1\,\Delta x_3\left[(\rho U_2 U_1)|_{x_2} - (\rho U_2 U_1)|_{x_2+\Delta x_2}\right] - \tau_{21}|_{x_2} + \tau_{21}|_{x_2+\Delta x_2}\\ &+ \Delta x_1\,\Delta x_2\left[(\rho U_3 U_1)|_{x_3} - (\rho U_3 U_1)|_{x_3+\Delta x_3}\right] - \tau_{31}|_{x_3} + \tau_{31}|_{x_3+\Delta x_3}\\ &+ \Delta x_2\,\Delta x_3\left[(P)|_{x_1} - (P)|_{x_1+\Delta x_1}\right] + \Delta x_1\,\Delta x_2\,\Delta x_3\,\rho g_1.\end{aligned} \tag{2.19}$$

From our definition of τ it is $-\tau$ that denotes the momentum that is transported into the fluid element at x. Equation (2.19) corresponds to a finite-volume formulation of the momentum equations for a hexahedral mesh aligned with the coordinate axis. Dividing by $\Delta x_1\ \Delta x_2\ \Delta x_3$ and taking the limit $\Delta x \to 0$ gives for all three components the Navier–Stokes equations

$$\frac{\partial U_1}{\partial t} + U_1\frac{\partial U_1}{\partial x_1} + U_2\frac{\partial U_1}{\partial x_2} + U_3\frac{\partial U_1}{\partial x_3} = -\frac{1}{\rho}\frac{\partial P}{\partial x_1} + \frac{1}{\rho}\frac{\partial \tau_{11}}{\partial x_1} + \frac{1}{\rho}\frac{\partial \tau_{21}}{\partial x_2} + \frac{1}{\rho}\frac{\partial \tau_{31}}{\partial x_3} + g_1,$$
$$\frac{\partial U_2}{\partial t} + U_1\frac{\partial U_2}{\partial x_1} + U_2\frac{\partial U_2}{\partial x_2} + U_3\frac{\partial U_2}{\partial x_3} = -\frac{1}{\rho}\frac{\partial P}{\partial x_2} + \frac{1}{\rho}\frac{\partial \tau_{12}}{\partial x_1} + \frac{1}{\rho}\frac{\partial \tau_{22}}{\partial x_2} + \frac{1}{\rho}\frac{\partial \tau_{32}}{\partial x_3} + g_2,$$
$$\frac{\partial U_3}{\partial t} + U_1\frac{\partial U_3}{\partial x_1} + U_2\frac{\partial U_3}{\partial x_2} + U_3\frac{\partial U_3}{\partial x_3} = -\frac{1}{\rho}\frac{\partial P}{\partial x_3} + \frac{1}{\rho}\frac{\partial \tau_{13}}{\partial x_1} + \frac{1}{\rho}\frac{\partial \tau_{23}}{\partial x_2} + \frac{1}{\rho}\frac{\partial \tau_{33}}{\partial x_3} + g_3. \tag{2.20}$$

These three equations can be rewritten as

$$\frac{\partial U_i}{\partial t} + \sum_j U_j\frac{\partial U_i}{\partial x_j} = -\frac{1}{\rho}\frac{\partial P}{\partial x_i} + \sum_j \frac{1}{\rho}\frac{\partial \tau_{ji}}{\partial x_j} + g_i.$$

Note that there is no summation over i, since i represents the three equations. In tensor notation these three equations are written

$$\frac{\partial U_i}{\partial t} + U_j\frac{\partial U_i}{\partial x_j} = -\frac{1}{\rho}\frac{\partial P}{\partial x_i} + \frac{1}{\rho}\frac{\partial \tau_{ji}}{\partial x_j} + g_i, \tag{2.21}$$

which for a Newtonian fluid becomes

$$\frac{\partial U_i}{\partial t} + U_j\frac{\partial U_i}{\partial x_j} = -\frac{1}{\rho}\frac{\partial P}{\partial x_i} + \nu\frac{\partial}{\partial x_j}\left(\frac{\partial U_i}{\partial x_j} + \frac{\partial U_j}{\partial x_i}\right) + g_i. \tag{2.22}$$

Equation (2.22) can be written in different forms since

$$\nu\frac{\partial}{\partial x_j}\left(\frac{\partial U_i}{\partial x_j} + \frac{\partial U_j}{\partial x_i}\right) = \nu\frac{\partial^2 U_i}{\partial x_j \partial x_j}$$

in incompressible flow with constant ρ and ν. In addition to gravity, there are additional external sources that may affect the acceleration of the fluid, e.g. electrical and magnetic fields. When reading Eq. (2.21) note that j should be summed over all dimensions, i.e. $j = 1$, 2 and 3, and i appears in all terms and for three dimensions, constituting the three equations as in Eq. (2.20).

Strictly it is the momentum equations that form the Navier–Stokes equations, but sometimes the continuity and the momentum equations together are called the Navier–Stokes equations. The Navier–Stokes equations are limited to macroscopic conditions. In reality the molecules move some distance before they collide, and the kinetic energy and consequently the velocity of the individual molecules are Boltzmann distributed. These effects must be accounted for at low pressures and in very small volumes. The Knudsen number relates the mean free path, λ, to the system dimension, $Kn = \lambda/L$. The average distance between collisions, i.e. the mean free path, in air at 1 atm and room temperature is $\sim$80 nm, and a correction of the Navier–Stokes equations and the boundary conditions is required for Knudsen numbers larger than ~0.02. Simulation of microfluids at dimensions below 5 μm at atmospheric pressure will require special boundary conditions [1].

2.4 Energy transport

Energy is present in many forms in flow, e.g. as kinetic energy due to the mass and velocity of the fluid, as thermal energy and as chemically bounded energy. We can then define the enthalpy as

$$\begin{aligned}
h &= h_m + h_T + h_C + \Phi && \text{total energy,}\\
h_m &= \tfrac{1}{2}\rho U_i U_i && \text{kinetic energy,}\\
h_T &= \sum_n m_n \int_{T_{\text{ref}}}^{T} c_{p,n}\,\mathrm{d}T && \text{thermal energy,}\\
h_C &= \sum_n m_n h_n && \text{chemical energy,}\\
\Phi &= \rho g_i x_i && \text{potential energy,}
\end{aligned}$$

where m_n is the mass fraction, $c_{p,n}$ the heat capacity and h_n the standard state enthalpy (heat of formation) for species n. The potential energy is often included in the kinetic energy.

The balance equation for total energy is

$$\frac{\partial h}{\partial t} = -\frac{\partial}{\partial x_j}\left[hU_j - k_{\text{eff}}\frac{\partial T}{\partial x_i} + \sum_n m_n h_n j_n - \tau_{kj}U_k\right] + S_h. \tag{2.23}$$

Here j_n is the diffusional flux of species n,

$$j_n = -D_n\frac{\partial C_n}{\partial x_j}.$$

The couplings between the energy equations and the momentum equations are weak for incompressible flows, and the equations for kinetic, thermal and chemical energies can be written separately.

2.4.1 The balance for kinetic energy

An equation for the kinetic energy including the potential energy can be deduced from the momentum equation by multiplying by velocity U_i:

$$U_i\frac{\partial U_i}{\partial t} + U_iU_j\frac{\partial U_i}{\partial x_j} = -\frac{1}{\rho}U_i\frac{\partial P}{\partial x_i} + \frac{U_i}{\rho}\frac{\partial \tau_{ij}}{\partial x_j} + U_i g_i. \tag{2.24}$$

By using

$$U_i \frac{\partial U_i}{\partial x_i} = \frac{1}{2} \frac{\partial U_i^2}{\partial x_i}$$

and defining $e \equiv \frac{1}{2}(U_1^2 + U_2^2 + U_3^2)$ [J kg^{-1} fluid] we obtain

$$\frac{\partial e}{\partial t} + U_j \frac{\partial e}{\partial x_j} = -\frac{1}{\rho} U_i \frac{\partial P}{\partial x_i} + \frac{1}{\rho} U_i \frac{\partial \tau_{ij}}{\partial x_j} + U_i g_i. \tag{2.25}$$

Note that this is a scalar equation and the expanded equation is obtained by summation over i and j. On multiplying by the density of the fluid and defining $h_m = \rho e$ [J m^{-3} fluid] and using the relations

$$\frac{\partial (PU_i)}{\partial x_i} = U_i \frac{\partial P}{\partial x_i} + P \frac{\partial U_i}{\partial x_i} \quad \text{and} \quad \frac{\partial (\tau_{ij} U_i)}{\partial x_i} = U_i \frac{\partial \tau_{ij}}{\partial x_i} + \tau_{ij} \frac{\partial U_i}{\partial x_i}$$

we obtain

$$\frac{\partial (h_m)}{\partial t} = -U_j \frac{\partial (h_m)}{\partial x_j} + P \frac{\partial U_i}{\partial x_i} - \frac{\partial (PU_i)}{\partial x_i} + \frac{\partial (\tau_{ij} U_i)}{\partial x_i} - \tau_{ij} \frac{\partial U_i}{\partial x_i} + \rho g U_i. \tag{2.26}$$

The accumulation and convection terms (the first two terms on the right-hand side) are straightforward and need no further comments. The work done by the gravity force (the sixth term on the right-hand side) is the change in potential energy due to gravity. Reversible conversion to heat (the third term on the right-hand side) stems from the thermodynamic cooling when a gas expands or heating when it is compressed. The work done by viscous forces (the fourth term on the right-hand side) is the accumulation of strain in some fluids, e.g. a rubber band. The irreversible conversion of kinetic energy into heat (the fifth term on the right-hand side) is, for Newtonian fluids,

$$\varepsilon = -\frac{1}{\rho} \tau_{ij} \frac{\partial U_i}{\partial x_j} = \frac{1}{2} \nu \left[\left(\frac{\partial U_i}{\partial x_j} + \frac{\partial U_j}{\partial x_i} \right) - \frac{2}{3} \frac{\partial U_i}{\partial x_i} \right]^2. \tag{2.27}$$

Owing to the fact that this is a squared term, the viscous dissipation term is always positive for Newtonian fluids. Heat is actually random movement of molecules or atoms, i.e. translational, rotational and vibrational movement. The difference between kinetic energy and heat is that kinetic energy has an average direction of movement whereas heat is random movement. With this perspective the dissipation term can be seen as viscous transport of fast molecules into areas with low average velocity and of slow molecules into areas with high average velocity. The molecules will collide with other molecules, transferring their average momentum, but will very soon lose their directional average and the movement is defined as heat.

The dissipation term is usually small and only with a very high velocity gradient will it be possible to measure the temperature increase due to viscous dissipation. In a stirred-tank reactor the power input by the impeller is of the order of 1 kW m^{-3}, which corresponds to a temperature increase of about 1 K h^{-1}. However, in turbulent flow this term becomes very important since it describes the decay of turbulence when the energy in the turbulent eddies is transferred into heat.

2.4.2 The balance for thermal energy

A balance for heat can be formulated generally by simply adding the source terms from the kinetic-energy balance and from chemical reactions:

$$\frac{\partial(\rho c_p T)}{\partial t} = -U_j \frac{\partial(\rho c_p T)}{\partial x_j} + k_{\text{eff}} \frac{\partial^2 T}{\partial x_j \partial x_j} - P \frac{\partial U_j}{\partial x_j} + \tau_{kj} \frac{\partial U_k}{\partial x_j} + \sum_m R_m(C, T)(-\Delta H_m) + S_T, \tag{2.28}$$

where the terms on the right-hand side are for accumulation, convection, conduction, expansion, dissipation and the reaction source.

Here the terms in the equation for transformation between thermal and kinetic energy, i.e. expansion and dissipation, occur as source terms. The relation to change in chemical energy is seen in the term for heat formation due to chemical reactions. Examples of the source term S_T are absorption and emission of radiation.

2.5 The balance for species

The balance for transport and reaction for species in constant-density fluids is described by

$$\frac{\partial C_n}{\partial t} + U_j \frac{\partial C_n}{\partial x_j} = \frac{\partial}{\partial x_j}\left(D_n \frac{\partial C_n}{\partial x_j}\right) + R(C, T) + S_n. \tag{2.29}$$

In most CFD programs, the concentration is replaced with the mass fraction

$$y_n = \frac{M_{v,n} C_n}{\rho}. \tag{2.30}$$

Transport and reaction will be discussed further in Chapter 5.

2.6 Boundary conditions

The 3D Navier–Stokes equations contain four dependent variables, U_1, U_2, U_3 and P. Depending on the conditions of the flow, we can define the boundary conditions in many different ways. The boundary conditions are just as important as the differential equations that determine the system, and the results of the simulations depend on the inlet and outlet conditions and the conditions at the walls as well as on the differential equations. We may also introduce boundary conditions due to simplifications of the computational domain, e.g. symmetry.

2.6.1 Inlet and outlet boundaries

The inlet velocity can be defined in terms of velocities or mass flow rate. In most CFD programs it is possible to enter the inlet condition as an average flow perpendicular to the surface or as a velocity-component distribution over the inlet surface. The way we define boundaries may affect the results, e.g. defining the inlet by an average velocity or by a parabolic laminar flow distribution will give the flow different momentum distributions and the total energy added by the inlet flow will be different in these two cases. An alternative to defining inlet velocities is the pressure inlet boundary condition that can be used when the inlet pressure is known without knowledge of the flow rate. The pressure inlet boundary condition is also useful when it is unknown whether the flow enters or exits at this position.

The standard outlet boundary condition is the zero-diffusion flux condition applied at outflow cells, which means that the conditions of the outflow plane are extrapolated from within the domain and have no impact on the upstream flow, i.e.

$$\phi|_{L-} = \phi|_{L+}. \tag{2.31}$$

The pressure outlet boundary condition is often the default condition used to define the static pressure at flow outlets. The use of a pressure outlet boundary condition instead of an outflow condition often results in a better rate of convergence when backflow occurs during iteration. Pressure outflow is also useful when there are several outflows. Specified outflow boundary conditions are used to model flow exits where the details of the inlet flow velocity and pressure are not known prior to solution of the flow problem. They are appropriate where the exit flow is close to a fully developed condition, since the standard outflow boundary condition assumes a normal gradient of zero for all flow variables except pressure.

Scalars, e.g. temperature and species, are usually defined as temperature and mass fractions in the inlet flow. Standard outlet conditions are usually the default.

2.6.2 Wall boundaries

The usual boundary condition for velocity at the walls is the ‘no-slip condition’, i.e. the relative velocity between the wall and the fluid is set to zero. For high-Reynolds-number turbulent flow the no-slip condition is still valid but the grid resolution is usually too coarse to specify the no-slip condition. In this case the velocity and shear close to the wall are modelled using a wall function. (See Chapter 4 for further details.) The no-slip condition may be inappropriate for non-Newtonian and multiphase flow and at large Knudsen numbers ($Kn > 0.02$), i.e. for low pressure or small dimensions.

For heat transfer, walls can be considered insulated or heat may be transferred through the walls. For heat there are several choices for boundary conditions, e.g. fixed heat flux, fixed temperature, convective heat transfer, radiation heat transfer or a combination of these boundary conditions. Heat transfer by radiation occurs mainly between solid surfaces, and the CFD program must be able also to calculate view angles in order to obtain accurate radiation boundary conditions.

The boundary conditions for species are usually termed 'no penetration', but diffusion to and reaction at the walls can occur. Evaporation and condensation are also possible wall boundary conditions.

2.6.3 Symmetry and axis boundary conditions

The time taken for simulations can be reduced significantly by using the geometrical symmetry of the problem. Mirror symmetry can halve or even further reduce the calculation region, and rotational symmetry, by defining a rotation axis, reduces a problem from 3D to 2D, which can decrease the time taken for simulation by several orders of magnitude. Symmetric initial and boundary conditions do not guarantee that the solution is symmetric, e.g. buoyancy-driven flows have a tendency to have several possible solutions depending on the initial conditions, and enforced symmetry conditions can produce erroneous results. Two-dimensional simulations may give very misleading results, e.g. the bubbles appearing in a simulation of a fluidized bed are cylinders in 2D and toroids in rotational symmetry. It should also be recognized that no net transport is allowed across a symmetry plane.

Periodic boundaries are convenient for DNS and LES simulations of turbulent flows since time-resolved inlet conditions are required. In periodic boundaries the inlet is set to the outlet. In this way the time-resolved solution will transfer the outlet solution to the inlet. However, to avoid having the boundary condition affect the simulation results, the residence time in the simulation domain should be long compared with the lifetime of the turbulent eddies. Periodic boundaries are also useful for rotating systems when only a fraction of the tangential direction is resolved, e.g. in simulation of turbines.

2.6.4 Initial conditions

Since the Navier–Stokes equations are nonlinear it is necessary to have an initial guess from which the solver can start the iterations. The better the initial conditions, the faster the final solution will convergence. It is also possible that the specified problem has multiple solutions and that the solution will converge to different solutions depending on the initial guess. It is always recommended that different initial conditions should be tested to evaluate convergence and to determine whether there are multiple stationary solutions. When multiple solutions are possible, the simulations must be transient, starting from correct initial conditions.

If a time-dependent solution is required, the actual initial conditions must be specified. Initial conditions for all variables that are to be solved must be specified. For example, in turbulence modelling the initial conditions for the variables describing the turbulence, e.g. the turbulence kinetic energy and the rate of dissipation, must also be set.

Owing to the numerical properties in transient simulations, it is in some cases more efficient to solve a steady-state case by transient simulations. In this case, we are not interested in accurate simulation of the transient behaviour, but only in obtaining a reliable steady-state solution. Exact initial conditions are not necessary, and it is possible to use larger time steps to obtain faster convergence.

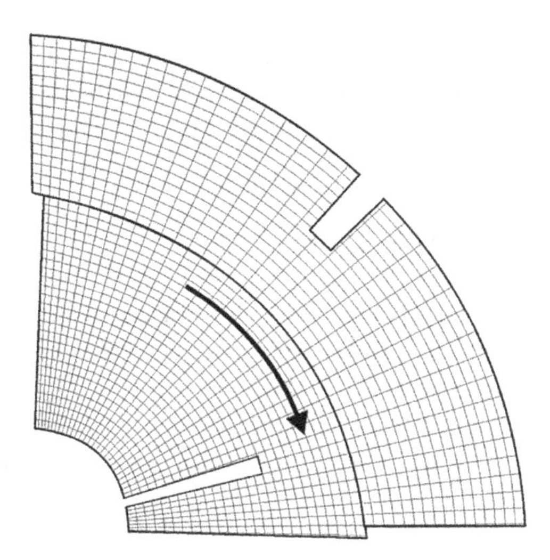

Figure 2.7 A mesh with a rotating and a stationary part.

2.6.5 Domain settings

There are occasions when we are interested only in relative velocities, e.g. flow through rotor blades in a compressor or flow in a centrifuge. We can then define the flow relative to the moving surfaces, and the CFD program will add volume forces due to acceleration, e.g. the centrifugal force.

It is also possible to have rotational parts together with stationary parts, e.g. in a stirred-tank reactor the impeller is rotating and the walls and baffles are stationary. We can then define a cylindrical volume around the impeller that is moving with the impeller and define the rest of the tank as stationary as shown in Figure 2.7.

In this model we do not have to specify the fluid velocities at the moving walls. Standard wall boundary conditions, e.g. no-slip conditions, are sufficient. By choosing the boundary between the moving and stationary parts at a point where the change in flow in the tangential direction is small, it is possible to obtain very good time-resolved flow simulations. The inner volume is stepped in time and the flow is distributed to the connecting cells in the outer volume. In some CFD programs it is also possible to use the average flow to obtain a steady-state solution. By proper choice of a rotating or sliding mesh it is even possible to obtain good average properties of a stirred-tank reactor.

2.7 Physical properties

Most physical properties of fluids, e.g. viscosity, density, diffusion, heat conductivity and surface tension, will vary locally due to variations in temperature, pressure and composition of the fluid. Algebraic equations describing these properties must be specified in the CFD program.

2.7.1 The equation of state

The relations among density, temperature, pressure and composition are described by the equation of state. For gases at low pressure the ideal-gas law can be used for compressible flows:

$$\rho = \frac{P}{RT \sum_n y_n / M_{w,n}}. \tag{2.32}$$

The ideal-gas law is also an option for incompressible flow if the pressure variation is moderate. The model will then correctly express the relationship between density and temperature required for e.g. natural convection problems.

For non-ideal gases many choices can be found in the literature, the most common of which are the law of corresponding states and the cubic equations of state. The law of corresponding states is defined as

$$Z = \frac{PV}{RT}, \tag{2.33}$$

where Z is a function of the reduced temperature and pressure. The cubic equations are in the form

$$P = \frac{RT}{V - b} - \frac{a}{V^2 + ubV + wb^2}, \tag{2.34}$$

where a, b, u and w are parameters. Depending on the parameters, they form van der Waals, Redlich–Kwong, Soave and Peng–Robinson equations of state.

For liquids the pressure dependence can often be neglected and a simple polynomial can describe the temperature dependence:

$$\rho = A + BT + CT^2 + DT^3 + \cdots. \tag{2.35}$$

2.7.2 Viscosity

At low pressure the viscosity increases slowly with temperature but there is only a small pressure dependence. The Chapman–Enskog theory provides an expression for the gas viscosity:

$$\mu_{\text{gas}} = \frac{5}{16} \frac{\sqrt{\pi m k_{\text{B}} T}}{\pi \sigma^2 \Omega T*}. \tag{2.36}$$

A more compressed form is Sutherland's law,

$$\mu = \frac{C_1 T^{3/2}}{T + C_2}, \tag{2.37}$$

where C_1 and C_2 are constants. For a multi-component system the total viscosity depends in a nonlinear fashion on the individual viscosities μ_i and mole fractions X_i:

$$\mu = \sum_i \frac{X_i \mu_i}{\sum_j X_i \phi_{ij}}, \tag{2.38}$$

where

$$\phi_{ij} = \left[1 + \left(\frac{\mu_i}{\mu_j}\right)^{1/2} \left(\frac{M_{w,j}}{M_{w,i}}\right)^{1/4}\right]^2 \Bigg/ \left[8\left(1 + \frac{M_{w,i}}{M_{w,j}}\right)\right]^{1/2}. \tag{2.39}$$

The viscosity increases with temperature for gases, whereas the viscosity decreases exponentially with temperature for liquids. The temperature dependence for liquid viscosity is often written as

$$\mu_{\text{liq}} = a\mathrm{e}^{b/T}. \tag{2.40}$$

For non-Newtonian fluids there are several models available. In this book we will cover only Newtonian fluids, but the interested reader can find additional theories in standard textbooks [2]. The standard models for turbulent flows assume Newtonian fluids, and empirical models are required for modelling turbulent viscosity.

Questions

(1) Why are diffusivity, kinematic viscosity and thermal diffusion similar in gases at low pressure?
(2) What is the molecular mechanism for viscous transport of momentum in gases?
(3) Why is it necessary to rewrite the continuity equation in CFD software?
(4) Why can a gas be treated as incompressible when there is a pressure drop?
(5) Why does viscous dissipation of kinetic energy form heat?
(6) What are standard outlet conditions?
(7) What is a no-slip condition at the wall and when can it be used?
(8) What is a periodic boundary condition?
(9) What are symmetry and axis boundary conditions?
(10) What is an equation of state?

3 Numerical aspects of CFD

This chapter introduces commonly used numerical methods. The aim is to explain the various methods so that the reader will be able to choose the appropriate method with which to perform CFD simulations. There is an extensive literature on numerical methods, and the interested reader can easily find textbooks. Appropriate references are [3, 4].

3.1 Introduction

In the previous chapter the physics of fluid flow has been presented. The concepts of how the flow is modelled, with and without turbulence, have been introduced. The Navier–Stokes, continuity and pressure equations have been derived, and model equations for some turbulence quantities will be discussed in following chapters. Expressions for heat and mass transfer have also been presented. The expressions for velocity, pressure, turbulence quantities and heat and mass transfer, together with the appropriate boundary conditions, constitute the core of the CFD problem.

So far there has been no discussion of how these equations are solved. This chapter will deal with the most fundamental aspects of numerical procedures for solving problems with CFD. The scope of the chapter is to give the reader a numerical background and an understanding of some of the numerical problems that can occur. Being aware of the existence of these problems and being able to avoid them are of crucial importance.

The *general transport equation for an arbitrary variable* ϕ in conservative form is now stated:

$$\rho \frac{\partial \phi}{\partial t} + \rho \frac{\partial (U_j \phi)}{\partial x_j} = \frac{\partial}{\partial x_j} \left(\Gamma \frac{\partial \phi}{\partial x_j} \right) + S_\phi. \tag{3.1}$$

This equation has been used in the previous chapter, e.g. the Navier–Stokes equations, Eq. (2.21), which actually are nothing other than transport equations for momentum (or rather velocity). The equation also appears as the transport equation for various turbulence quantities in Chapter 4. Unfortunately, it is generally not possible to solve equations of this type analytically, since they are nonlinear and often contain both spatial and temporal derivatives. This requires the application of numerical methods.

3.2 Numerical methods for CFD

The pioneers of CFD employed finite differences to approximate the governing equations describing fluid mechanics. With finite differences, the partial spatial and temporal derivatives appearing in the equations are approximated through Taylor series. Although there is no formal restriction, finite differences are typically employed only on Cartesian geometries. Since most of the problems engineers tackle do not take place in a square box, finite differences are not often used for practical problems.

Finite-element methods require a 2D or 3D mesh and are very flexible in terms of geometry and mesh elements; almost any type of mesh element can be employed. At each mesh element, a base function is used. This base function should locally describe the solution of (part of) the governing equation to be approximated. The finite-element method aims to minimize the difference between the exact solution and the collection of base functions; this can be done e.g. by a Galerkin method. There is no dispute that finite-element methods are the preferred method of choice for solid-mechanics problems.

However, problems in the fluid-mechanics area are generally governed by *local* conservation. For instance, the continuity equation dictates the local conservation of mass. Local conservation is not necessarily a property of the finite-element method, since the difference between the base functions and the exact solution is minimized globally. The adaptation of the finite-element method to reflect local conservation is still very much the focus of numerical research, therefore the method has historically not been used as much for CFD.

3.2.1 The finite-volume method

The principle of the finite-volume method is local conservation, and this is the key reason for its success in CFD. To solve the equations numerically with the finite-volume method, the entire computational domain is divided into 'small' sub-volumes, so-called *cells*. Employing Gauss' law, the partial derivatives expressing a conservation principle, such as div u, can be rewritten at each cell as an algebraic contribution. The governing equation, expressed in the partial differential equations, is reformulated, at each computational cell, into a set of linear algebraic equations

Usually, these equations are solved numerically in an iterative manner. The price for this so-called *discretization* of the domain is the introduction of a numerical error into the solution. It is important to control the magnitude of the error after a solution has been obtained, and, since it can be shown that this error vanishes as the cell size approaches zero, a sufficient decrease of the cell size will often reduce the error well enough. On the other hand, reducing the cell size too much will create an unnecessarily large number of cells, which will increase the computational effort required and possibly yield prohibitive simulation times. Finding a fast but still accurate way of solving the CFD problem is one of the CFD engineer's most important tasks.

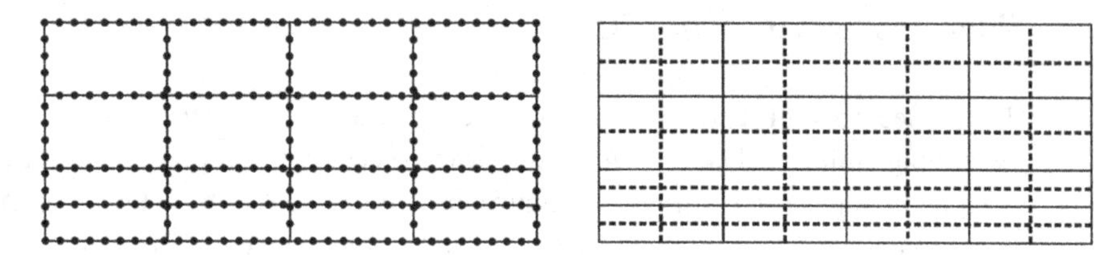

Figure 3.1 Control volumes. The cell-centred algorithm (left) and the node-centred algorithm (right). The mesh elements are depicted with thin lines, and the control volumes with dashed lines.

3.2.2 Geometrical definitions

Now some definitions have to be made. The *cell* has already been defined. Each cell is surrounded by its *faces*. These faces form a grid pattern throughout the domain. A grid that contains only cells with all internal angles equal to 90° is called a *structured grid*, and this is the type of grid that will be dealt with in this chapter. Since a structured grid requires that the physical geometry itself must be rectangular, this type of grid is not very common in reality. Many industrial cases contain parts of complex geometry that cannot be divided into purely rectangular cells. Grids can be created in 1D, 2D or 3D, depending on the number of computational dimensions. A problem can often be regarded as being of 1D or 2D even though all fluid flows are 3D in 'reality'.

In the work of defining the grid it must also be taken into account whether the solver is to use a cell-centred or a node-centred algorithm. A cell-centred solver algorithm creates control volumes that are completely identical to the grid. A node-centred solver creates its control volumes around the grid nodes instead. A grid node is situated at each intersection of cell edges, see also Figure 3.1. Even though choosing the algorithm is primarily a solver issue, the choice has to be made already during the stage of grid creation. This follows as a consequence of the fact that the two algorithms put different demands on the grid, especially in near-wall regions.

3.3 Cell balancing

Integrating the general transport equation, Eq. (3.1), over a control volume (c.v.) yields

$$\int_{\mathrm{c.v.}} \rho \frac{\partial \phi}{\partial t}\,\mathrm{d}V + \int_{\mathrm{c.v.}} \rho \frac{\partial (U_j \phi)}{\partial x_j}\,\mathrm{d}V = \int_{\mathrm{c.v.}} \frac{\partial}{\partial x_j}\left(\Gamma \frac{\partial \phi}{\partial x_j}\right)\mathrm{d}V + \int_{\mathrm{c.v.}} S_\phi\,\mathrm{d}V. \tag{3.2}$$

For the time being, only steady problems are considered. Thus, the accumulation term will be zero. Transient problems (unsteady problems) are studied in later sections in this chapter.

The next step in solving a problem with finite volumes is to reformulate Eq. (3.2) in algebraic form. This requires the elimination of all integral signs and derivatives. To do

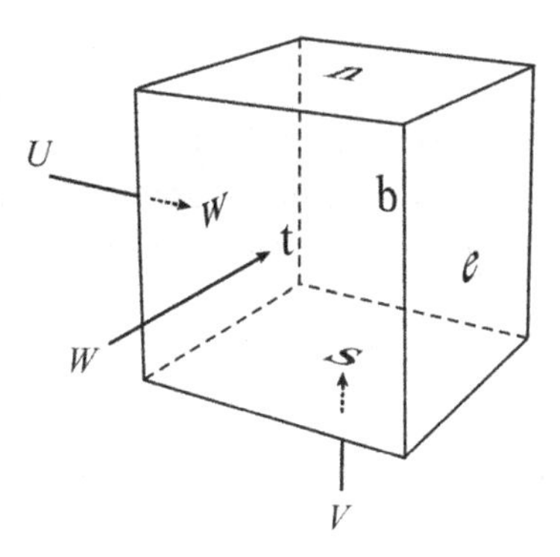

Figure 3.2 The faces of a control volume.

this, some approximations have to be introduced. The following section will discuss the process in detail.

3.3.1 The convective term

The terms in Eq. (3.2) are now studied one by one, starting with the convective term

$$\int_{\text{c.v.}} \rho \frac{\partial (U_j \phi)}{\partial x_j} \, \mathrm{d}V. \tag{3.3}$$

This term represents the *net* flux of ϕ transported out of the cell by convection, i.e. transported with the flow. The flow can enter or leave the cell only through any of its faces. According to Gauss' theorem it is possible to rewrite Eq. (3.3) as

$$\int_{\text{c.s.}} \rho U_j n \phi \, \mathrm{d}A, \tag{3.4}$$

where c.s. denotes the control-volume surfaces, i.e. the faces that surround the cell, and $\mathrm{d}A$ is the area that surrounds the volume $\mathrm{d}V$. Here n is a normal vector pointing outwards from $\mathrm{d}A$. The product $U_j n$ is thus the velocity perpendicular to the surface $\mathrm{d}A$. In a structured grid one can evaluate Eq. (3.4) into

$$-\rho \left[(AU\phi)_{\mathrm{w}} - (AU\phi)_{\mathrm{e}} + (AV\phi)_{\mathrm{s}} - (AV\phi)_{\mathrm{n}} + (AW\phi)_{\mathrm{t}} - (AW\phi)_{\mathrm{b}} \right]. \tag{3.5}$$

Here, A is the appropriate face area and U, V and W are the velocities in the x, y and z directions, respectively. The indices w, e, s, n, b and t denote the west, east, south, north, bottom and top cell faces. The term $(AV\phi)_{\mathrm{n}}$ thus takes into account the flux of ϕ through the northern cell face that has size $\Delta x \Delta z$. See also Figure 3.2.

The negative signs in Eq. (3.5) come from the definition of west, east etc. in relation to the coordinate axis. For example, n equals -1 at the west surface, giving that the flux $(U\phi)_{\mathrm{w}}$ is negative if U is positive there. If Eq. (3.5) is regarded as an integral mass balance, it makes sense in the way that what goes into the cell also comes out for source-free, steady conditions when there is no other means of transport than convection.

An apparent problem when evaluating the convective term is that the face values of U_j and ϕ must be known. As mentioned earlier, the transport equations are not solved on the faces. Several approaches to overcome this will be discussed later in the section about discretizing schemes.

3.3.2 The diffusion term

Next, a closer look is taken at the diffusion term,

$$\int_{\text{c.v.}} \frac{\partial}{\partial x_j}\left(\Gamma \frac{\partial \phi}{\partial x_j}\right) \mathrm{d}V. \tag{3.6}$$

The diffusion term takes into account the transport of ϕ by diffusion. Equation (3.6) can be treated in a similar way to the convective term, Eq. (3.3). By making use of Gauss' theorem,

$$\int_{\text{c.s.}} \Gamma \frac{\partial \phi}{\partial x_j} n \, \mathrm{d}A. \tag{3.7}$$

Using the same notation as in the convective case, Eq. (3.7) can be evaluated to give a similar expression:

$$-\left[\left(A\Gamma\frac{\partial \phi}{\partial x}\right)_{\text{w}} - \left(A\Gamma\frac{\partial \phi}{\partial x}\right)_{\text{e}} + \left(A\Gamma\frac{\partial \phi}{\partial y}\right)_{\text{s}} - \left(A\Gamma\frac{\partial \phi}{\partial y}\right)_{\text{n}} + \left(A\Gamma\frac{\partial \phi}{\partial z}\right)_{\text{t}} \left(A\Gamma\frac{\partial \phi}{\partial z}\right)_{\text{b}}\right]. \tag{3.8}$$

3.3.3 The source term

The last term in the general transport equation is the source term,

$$\int_{\text{c.v.}} S_\phi \, \mathrm{d}V. \tag{3.9}$$

The source term takes into account any generation or dissipation of ϕ. The body force due to gravity in the Navier–Stokes equation for the y momentum is an example of a source term as discussed previously. The pressure gradient term in the Navier–Stokes equation is another example. To be able to rewrite Eq. (3.9) without using any integral signs, simply take a cell mean value of S_ϕ and move it outside the integral sign. Thus,

$$\int_{\text{c.v.}} S_\phi \, \mathrm{d}V \approx \overline{S}_\phi V. \tag{3.10}$$

Here, $\overline{S}_\phi$ is the mean value of S_ϕ in the cell.

The original transport equation, Eq. (3.1), is now transformed into equations that can be solved algebraically, Eqs. (3.5), (3.8) and (3.10). It has also been concluded

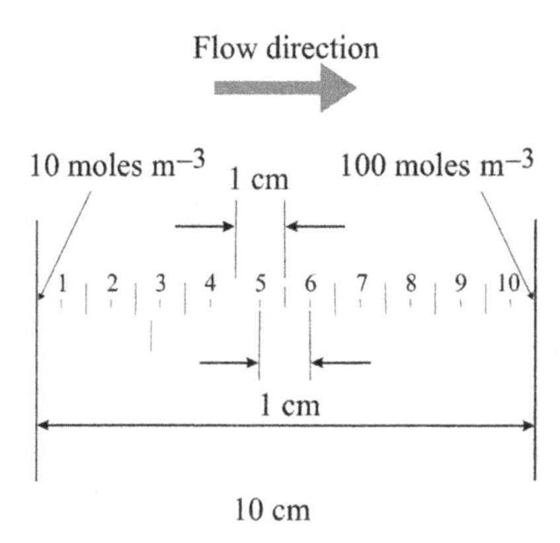

Figure 3.3 The experimental set-up in Example 1.

that the face values of ϕ, Γ and U_j need to be predicted, and the gradient of ϕ at the faces is needed. If a source is present, the correct mean values of the sources in all cells must be created. It must also be remembered that there can be many different scalars, and that they are usually dependent on each other, forcing us to solve their equations simultaneously in each cell. Hence, the set of equations often becomes nonlinear.

Equation (3.5), (3.8) and (3.10) will now be used to solve an easy and clarifying 1D problem, Example 1.

3.4 Example 1 – 1D mass diffusion in a flowing gas

The flow of an inert gas I between two perforated surfaces is simulated. The distance between the surfaces is 10 cm, and it is assumed that the surface areas are infinite. This assumption is made so that the problem can be treated as being 1D. The gas velocity is 1 mm s^{-1}. At the west surface the species concentration is 10 moles m^{-3} of A and at the east surface the concentration is 100 moles m^{-3} of A. Assuming that the total density is 1 kg m^{-3}, the diffusion constant for A in I is 10^{-4} m^2 s^{-1}. The objective is to predict the profile of A between the surfaces, using ten equidistant cells. See Figure 3.3 for the simulation set-up.

3.4.1 Solution

This is a 1D problem. The general transport equation, Eq. (3.1), can be evaluated according to the previous sections with a somewhat simpler expression than for the general 3D case. By way of illustration, the steps are presented. Here, ϕ denotes the molar concentration of A.

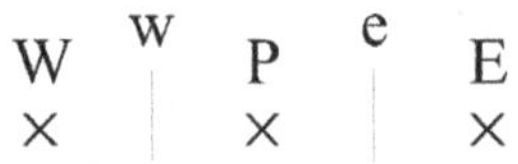

$$\begin{array}{ccccc} & \text{w} & & \text{e} & \\ \text{W} & & \text{P} & & \text{E} \\ \times & | & \times & | & \times \end{array}$$

Figure 3.4 Neighbouring cells and cell faces.

Integrating Eq. (3.1) over a 1D control volume forms

$$\int_{\mathrm{w}}^{\mathrm{e}} \rho \frac{\partial \phi}{\partial t}\,\mathrm{d}x + \int_{\mathrm{w}}^{\mathrm{e}} \rho \frac{\partial (U\phi)}{\partial x}\,\mathrm{d}x = \int_{\mathrm{w}}^{\mathrm{e}} \frac{\partial}{\partial x}\left(\Gamma \frac{\partial \phi}{\partial x}\right)\mathrm{d}x + \int_{\mathrm{w}}^{\mathrm{e}} S_\phi\,\mathrm{d}x. \tag{3.11}$$

Here $\Gamma = \rho D$, where D is the diffusion constant and ρ the total density.

Compared with Eq. (3.2) for the 3D case, it can be seen that the 3D control volumes have been replaced with 1D control volumes. This simply requires integration from the west face to the east instead of over the 3D volume. As shown earlier, it is assumed that steady conditions prevail and thereby it is possible to neglect the accumulation term. Further, there is no internal source of species A in our system. This results in the reduced equation

$$\int_{\mathrm{w}}^{\mathrm{e}} \rho \frac{\mathrm{d}(U\phi)}{\mathrm{d}x}\,\mathrm{d}x = \int_{\mathrm{w}}^{\mathrm{e}} \frac{\mathrm{d}}{\mathrm{d}x}\left(\Gamma \frac{\mathrm{d}\phi}{\mathrm{d}x}\right)\mathrm{d}x.$$

This can be evaluated into the algebraic form

$$\left[(\rho U\phi)_{\mathrm{e}} - (\rho U\phi)_{\mathrm{w}}\right] = \left[\left(\Gamma \frac{\mathrm{d}\phi}{\mathrm{d}x}\right)_{\mathrm{e}} - \left(\Gamma \frac{\mathrm{d}\phi}{\mathrm{d}x}\right)_{\mathrm{w}}\right]. \tag{3.12}$$

This equation holds for all cells. To proceed, estimates are required for the face values of ϕ and U and the gradient of ϕ at the faces. This can be done in many ways, but the most straightforward solution is to use a linear interpolation from neighbouring cells. Starting with the face values of ϕ,

$$\begin{aligned} \phi_{\mathrm{w}} &= \frac{\phi_{\mathrm{W}} + \phi_{\mathrm{P}}}{2}, \\ \phi_{\mathrm{e}} &= \frac{\phi_{\mathrm{P}} + \phi_{\mathrm{E}}}{2}. \end{aligned} \tag{3.13}$$

Here, W denotes the western neighbour cell, E the eastern neighbour cell and P the present cell (w and e are the face values as defined earlier). See also Figure 3.4. Be aware that Eq. (3.13) is merely an approximation and through this a numerical error is introduced into the solution. U is constant at 1 mm s^{-1} so the face values of U are already known. To estimate the gradient of ϕ at the faces, the following equations are used:

$$\begin{aligned} \left(\frac{\mathrm{d}\phi}{\mathrm{d}x}\right)_{\mathrm{w}} &= \frac{\phi_{\mathrm{P}} - \phi_{\mathrm{W}}}{x_{\mathrm{P}} - x_{\mathrm{W}}}, \\ \left(\frac{\mathrm{d}\phi}{\mathrm{d}x}\right)_{\mathrm{e}} &= \frac{\phi_{\mathrm{E}} - \phi_{\mathrm{P}}}{x_{\mathrm{E}} - x_{\mathrm{P}}} \end{aligned} \tag{3.14}$$

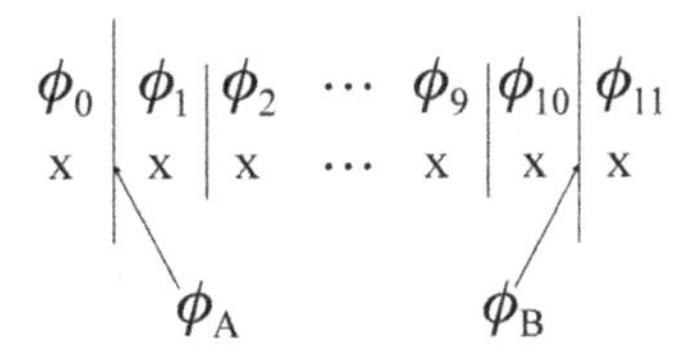

Figure 3.5 The cells. The shadow cells 0 and 11 are situated outside the physical domain. ϕ_A and ϕ_B are the boundary values at the western and eastern physical surfaces, respectively.

This is a first-order Taylor approximation and x_E, x_W and x_P are the cell coordinates of the eastern, western and present cells, respectively. Also, by using Eq. (3.14) a numerical error is introduced.

Equations (3.12), (3.13) and (3.14) can now be combined, which results in

$$\left[\left(\rho U \frac{\phi_P + \phi_E}{2}\right) - \left(\rho U \frac{\phi_P + \phi_W}{2}\right)\right] = \left[\left(\Gamma \frac{\phi_E - \phi_P}{x_E - x_P}\right) - \left(\Gamma \frac{\phi_P - \phi_W}{x_P - x_W}\right)\right]. \quad (3.15)$$

Equation (3.15) can be solved once for each cell, giving a total of ten equations (one per cell). Each equation will contain three unknowns, ϕ_W, ϕ_E and ϕ_P. However, the unknown value of, for example, ϕ_E in the equation for cell number 4 will come back as ϕ_P on solving Eq. (3.15) for cell number 5. Thus, the total number of unknown variables will be 12; one ϕ per cell plus two extra ϕ values for the farmost eastern and farmost western surfaces, i.e. the boundaries. Numerical values for the boundaries are given in the assignment, giving ten equations and ten unknown variables. The equation system can thus be solved and the profile determined.

Before solving the equation system, Eq. (3.15) is re-arranged in a slightly different form:

$$\underbrace{\left(\frac{\Gamma}{x_E - x_P} + \frac{\Gamma}{x_P - x_W}\right)}_{B}\phi_P + \underbrace{\left(\frac{\rho U}{2} - \frac{\Gamma}{x_E - x_P}\right)}_{C}\phi_E + \underbrace{\left(-\frac{\rho U}{2} - \frac{\Gamma}{x_P - x_W}\right)}_{A}\phi_W = 0. \quad (3.16)$$

Cells 2–9 are identical in the sense that the distances to the neighbouring cells are identical. For the cells that are closest to the boundaries, i.e. cells 1 and 10, there are no cells directly to the west and east, respectively. These cells thus require extra attention. A way to avoid having to treat these two cells in a special way is to create two 'imaginary' cells, so-called *shadow cells*, outside the computational domain. If the shadow cells are placed so that the distance from them to the closest real cell is the same as the distance between two real cells, then cells 1 and 10 are not treated any differently. This will make the solution procedure easier. See Figure 3.5 for more details. The shadow cells are called ϕ_0 and ϕ_{11}, respectively, and the boundary values are called ϕ_A and ϕ_B. Using linear interpolation, the shadow cell is calculated from $\phi_A = (\phi_0 + \phi_1)/2$.

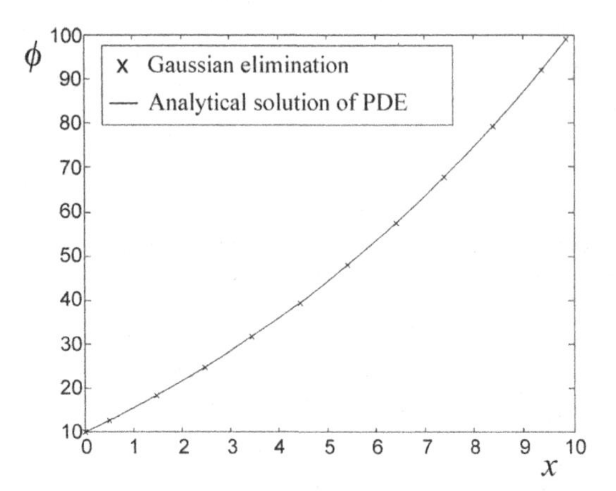

Figure 3.6 A plot of molar concentration as a function of distance from the left plate solved with Gaussian elimination. The exact solution of the PDE has been added for comparison.

Ten identical equations are now arranged for cells 1–10 and two slightly modified ones for the shadow cells. In matrix form,

$$\begin{pmatrix} 1 & 1 & 0 & 0 & 0 & 0 & 0 & 0 & 0 & 0 & 0 & 0 \\ A & B & C & 0 & 0 & 0 & 0 & 0 & 0 & 0 & 0 & 0 \\ 0 & A & B & C & 0 & 0 & 0 & 0 & 0 & 0 & 0 & 0 \\ 0 & 0 & A & B & C & 0 & 0 & 0 & 0 & 0 & 0 & 0 \\ 0 & 0 & 0 & A & B & C & 0 & 0 & 0 & 0 & 0 & 0 \\ 0 & 0 & 0 & 0 & A & B & C & 0 & 0 & 0 & 0 & 0 \\ 0 & 0 & 0 & 0 & 0 & A & B & C & 0 & 0 & 0 & 0 \\ 0 & 0 & 0 & 0 & 0 & 0 & A & B & C & 0 & 0 & 0 \\ 0 & 0 & 0 & 0 & 0 & 0 & 0 & A & B & C & 0 & 0 \\ 0 & 0 & 0 & 0 & 0 & 0 & 0 & 0 & A & B & C & 0 \\ 0 & 0 & 0 & 0 & 0 & 0 & 0 & 0 & 0 & A & B & C \\ 0 & 0 & 0 & 0 & 0 & 0 & 0 & 0 & 0 & 0 & 1 & 1 \end{pmatrix} \begin{pmatrix} \phi_0 \\ \phi_1 \\ \phi_2 \\ \phi_3 \\ \phi_4 \\ \phi_5 \\ \phi_6 \\ \phi_7 \\ \phi_8 \\ \phi_9 \\ \phi_{10} \\ \phi_{11} \end{pmatrix} = \begin{pmatrix} 2\phi_{\mathrm{A}} \\ 0 \\ 0 \\ 0 \\ 0 \\ 0 \\ 0 \\ 0 \\ 0 \\ 0 \\ 0 \\ 2\phi_{\mathrm{B}} \end{pmatrix}, \quad (3.17)$$

where the constants A, B and C are defined in Eq. (3.16). Since the coefficients A, B and C are not functions of ϕ_i, it is possible to solve Eq. (3.17) analytically with Gaussian elimination. This gives the following numerical values of ϕ_i at the cells: $\phi_0 = 7.38$, $\phi_1 = 12.61$, $\phi_2 = 18.39$, $\phi_3 = 24.78$, $\phi_4 = 31.84$, $\phi_5 = 39.65$, $\phi_6 = 48.28$, $\phi_7 = 57.81$, $\phi_8 = 68.35$, $\phi_9 = 80.00$, $\phi_{10} = 92.88$ and $\phi_{11} = 107.11$.

Remember that cells 0 and 11 are shadow cells and are used only to make the solution procedure easier. They are listed here for paedagogical reasons. For a plot of the results, please refer to Figure 3.6. It can be observed that there is only a very small deviation from the exact solution of Eq. (3.11) even though the approximations in Eq. (3.13) and Eq. (3.14) have been introduced.

3.4.2 Concluding remarks

The solution strategy for a simple CFD problem has now been demonstrated. However, it must be kept in mind that the solved example contained many simplifications and was well defined in many ways. Generally, it is not possible to solve a CFD problem with a *direct* method, such as Gaussian elimination. The focus is then turned towards *iterative* methods.

Some examples of simplifications in the solved Example 1 are the following.

- The problem could be treated as 1D due to symmetries. A problem in 2D or 3D would, of course, generate more cells; in this case a 3D treatment would give 1000 cells instead of 10, assuming that the grid density was kept constant and the computational domain had a cubic geometry. The cells were placed with constant spacing, generating a so-called *equidistant* grid.
- Further, the presence of a constant velocity made the solution process easier. Usually, the velocities cannot be predetermined, and thus one requires a solution to a transport equation for each velocity component as well. If this is the case, the equation system, Eq. (3.16), will contain nonlinear terms like $\rho U\phi$, where both U and ϕ are variables. This means that matrices can no longer be created with constant coefficients like in Eq. (3.17). The equation system must then be solved by some iterative technique, e.g. the *Gauss–Seidel* method (see below for more details).
- Another simplification introduced was the linear approximations of the variables and the gradients at the faces. Obviously, at high positive flow rates, the face values between, for example, cell 1 and cell 2 must be more dependent on cell 1 than on cell 2. Thus, linear interpolation, or *central differencing*, must be used with caution.
- The fluid properties were assumed to be constant. In non-isothermal situations, the temperature must be calculated in order for these entities to be predicted. A transport equation for temperature has to be solved. However, in order to solve the energy equation, the fluid properties must be known. This requires iterative methods.

Questions

(1) Propose a general expression for the transport of the entity ϕ. Give a physical interpretation and units to the terms.
(2) What is discretization? Is it always a necessary step in the solving of a CFD problem?
(3) How many boundary conditions are required in order to solve a steady diffusion–convection problem in n computational dimensions? Does this number change if diffusion is neglected? Give a physical explanation!
(4) What sources of error were introduced in Example 1?

3.5 The Gauss–Seidel algorithm

As discussed earlier, iterative methods are always used in CFD. Most commercial CFD codes use some variant of the Gauss–Seidel algorithm (GSA). Some basic knowledge of the GSA makes it easier to understand how these codes work and also what problems can occur.

For demonstration, Example 1 will be solved again, this time with the GSA. First, some general words about the GSA are in order. Starting with Eq. (3.16), it is possible to isolate ϕ_P from the equation:

$$\phi_P = \frac{a_E\phi_E + a_W\phi_W}{a_P}, \tag{3.18}$$

where

$$\begin{aligned} a_E &= \left(-\frac{\rho U}{2} + \frac{\Gamma}{x_E - x_P}\right), \\ a_W &= \left(\frac{\rho U}{2} + \frac{\Gamma}{x_P - x_W}\right), \\ a_P &= \left(\frac{\Gamma}{x_E - x_P} + \frac{\Gamma}{x_P - x_W}\right). \end{aligned} \tag{3.19}$$

The main difference between using the GSA and Gaussian elimination is that the GSA is an *iterative* method. Equation (3.18) is solved for each cell in an iterative manner. Start with cell 1; to be able to solve for this cell, the numerical value of the variable in the shadow cell, ϕ_0, and also the numerical value in cell 2, ϕ_2, are needed. Since there is no numerical value of ϕ in any cell, a starting guess, an *initialization*, for all ϕ_i must be made. ϕ_1 is then solved for. Shifting our attention to ϕ_2, Eq. (3.18) is solved for this cell. In the calculation of ϕ_2, the calculated value of ϕ_1 and the starting guess for ϕ_3 are used. Thus, the GSA uses the values calculated already in the same iteration sweep, which makes the solution converge faster. The procedure is then carried out until ϕ_i in all cells have been calculated. But, there is a problem here; since ϕ_1 was calculated as a function of ϕ_0 and ϕ_2, the equation for ϕ_1 is no longer satisfied since a new value of ϕ_2 has been calculated. Thus, the procedure must be repeated many times; i.e. it requires *iteration*. This is done until *convergence* is reached, i.e. the numerical values of ϕ_i change by less than a specified threshold amount set by the user. Discussion of how this is usually done in practice will come later.

3.6 Example 2 – Gauss–Seidel

This time, Example 1 will be solved numerically using the GSA. The procedure given in Figure 3.7 will be followed.

On looking at the expressions for a_W, a_E and a_P, one may see that they are identical for all cells and do not change during the iterations. We use the numerical values given in the assignment, $a_E = 0.0095$, $a_W = 0.0105$ and $a_P = 0.0200$, for all cells. A high numerical value of e.g. a_E means that cell (P) is highly influenced by it, whereas a low number means the opposite; see Eq. (3.18). As can be seen here, a_W is larger than a_E, meaning that the value for cell P is more influenced by its western neighbour than by its eastern one. This shouldn't come as any surprise, knowing that the gas is flowing from west to east and, thus, the western cell should play a more dominant role in the calculation of cell P.

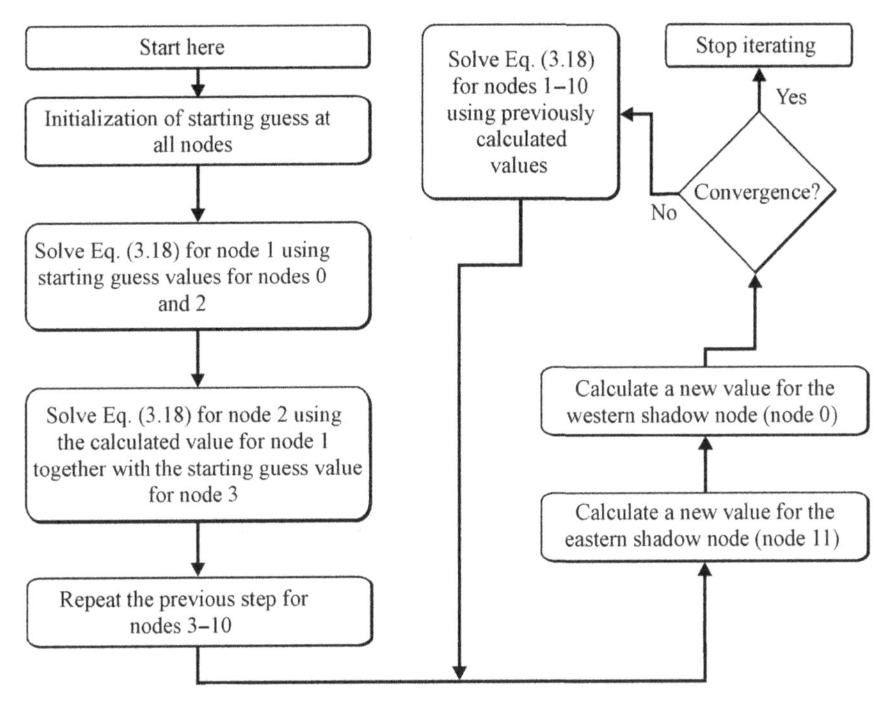

Figure 3.7 The Gauss–Seidel solving procedure for Example 2.

Next, starting values for all cells are initialized. For simplicity, set all ϕ_i equal to 50 moles m^{-3}. Making a 'good' starting guess is not that important in this example, but in more complex problems the starting guess can be vital even to reach convergence. This will be discussed further in the section about unsteady problems. Special attention is required for the shadow cells. As discussed earlier, it was through these cells that the boundary conditions were introduced. Given the numerical value of the boundaries, the numerical value of the shadow cells can be determined.

$$\phi_0 = 2\phi_A - \phi_1 = -30,$$
$$\phi_{11} = 2\phi_B - \phi_{10} = 150.$$

Now, Eq. (3.18) can be solved for all cells, beginning with cell 1 (see also Figure 3.7):

$$\phi_1 = \frac{0.0095 \times 50 + 0.0105 \times (-30)}{0.0200} = 8,$$

$$\phi_2 = \frac{0.0095 \times 50 + 0.0105 \times 8}{0.0200} = 27.95,$$

$$\phi_3 = \frac{0.0095 \times 50 + 0.0105 \times 27.95}{0.0200} = 38.42,$$

$$\ldots$$

$$\phi_9 = \frac{0.0095 \times 50 + 0.0105 \times 49.53}{0.0200} = 49.75,$$

$$\phi_{10} = \frac{0.0095 \times 150 + 0.0105 \times 49.75}{0.0200} = 97.37.$$

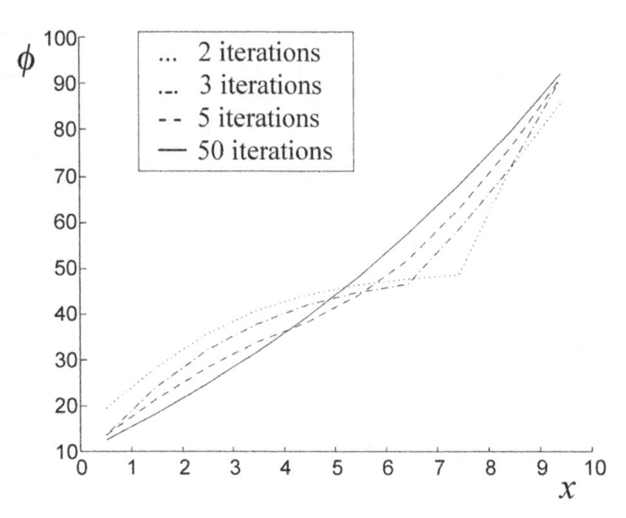

Figure 3.8 A plot of molar concentration as a function of the distance from the left plate at different stages in the iterative process.

On looking at the numerical values of the cells, a tendency for the low value at the western boundary to 'spread' through the domain can be seen. This is because iterations are sweeping from west to east, not vice versa. Iterating from east to west would have resulted in a profile, at this stage, with cell values higher than 50 for almost all cells. In the present iteration, information is transported from west to east, from low cell numbers to higher cell numbers. At convergence, the mode of iteration should not have any impact on the result, but it often affects the computational time that is required for convergence.

Additional iterations are run before continuing the discussion. For iteration 40 $\phi_1 = 12.6027$, $\phi_2 = 18.3587$, $\phi_3 = 24.7271$, $\phi_4 = 31.7741$, $\phi_5 = 39.5721$, $\phi_6 = 48.2000$, $\phi_7 = 57.7443$, $\phi_8 = 68.2998$, $\phi_9 = 79.9706$ and $\phi_{10} = 92.8714$. See Figure 3.8.

From the numerical data produced during the iterative process, it is possible to see that the rate of change of a numerical value at a specific cell decreases with iteration. This is also shown in Figure 3.9, where all cell values have been plotted against the number of iterations. Eventually, after continuing iterations for a 'very long' time, the same solution as was obtained in Example 1 will be reached, but already after a few iterations this solution is very close. Notice that there have been no additional sources of error introduced into the equations apart from the previously discussed discretization error and the error stemming from face-value or face-gradient approximations. However, if the iterations are interrupted 'too soon', the solution obtained can be far from the 'true numerical' solution. Thus, knowing when to stop iterating is important. In previous sections, the concept of convergence was briefly discussed; and it was stated that a solution is converged when it's 'close enough' to the solution to the original set of partial differential equations (PDEs). A proper measure of convergence will now be discussed.

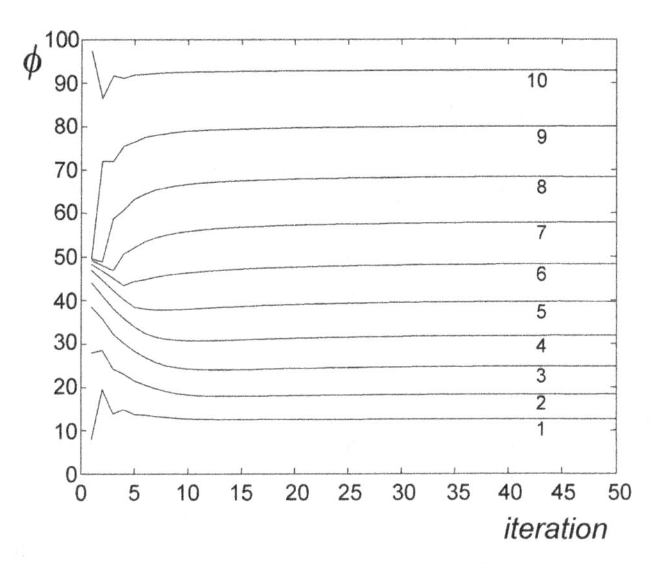

Figure 3.9 A plot of molar concentration at the cells during iterations.

3.7 Measures of convergence

The objective with measuring convergence is to know when to stop a CFD simulation. The code should iterate until a criterion is fulfilled. It should then stop. There are several different approaches that can be employed to find out when a solution has converged. Since the exact solution to the set of PDEs is generally unknown (otherwise there would be no need for any iterations), it is not possible to compare the numerical solution with this exact solution. Other methods must be found. Maybe the easiest one is to state that 'when no cell value differs by more than a small threshold, for example 0.01, from iteration to iteration, the solution has converged'. In Example 2, this would probably be quite a good measure. On the other hand, if the boundary values were very small, like 0.001 for the western boundary and 0.002 for the eastern boundary, then the solution would probably have converged before we'd even started iterating. The cell values would not change enough in an absolute sense. So just looking at the absolute difference

$$\left|\phi_i^{\text{new}} - \phi_i^{\text{old}}\right| < \varepsilon_{\text{thres}}$$

is not always an appropriate measure.

A similar approach would instead be to study the relative change

$$\left|\frac{\phi_i^{\text{new}} - \phi_i^{\text{old}}}{\phi_i^{\text{old}}}\right| < \varepsilon_{\text{thres}}.$$

A problem with this approach can occur if the cell values are very close to zero, since then the relative change would always be very large, and numerical problems can arise. However, this approach is used quite often, primarily because it's relatively easy to implement in a code and it turns out to be quite good.

A very common approach is to calculate the error in Eq. (3.18),

$$R = \sum_{\text{all cells}} |a_P\phi_P - a_W\phi_W - a_E\phi_E|. \tag{3.20}$$

If this error is scaled with an appropriate factor, a dimensionless number on which one can impose a criterion results. For example, if F is an appropriate scaling factor,

$$\frac{R}{F} < \varepsilon_{\text{thres}}, \qquad \text{e.g.} \qquad F = \sum_{\text{all cells}} |a_p\phi_p| + |a_W\phi_W| + |a_E\phi_E|.$$

In some commercial CFD software the scaling factor is simply set to the highest unscaled residual obtained within the first iterations. Hence, a bad starting guess makes it easier to reach convergence, and vice versa.

A completely different way to measure convergence is to check whether the domain upholds conservativeness. In the previous examples one could check the amount of A being transported into the domain and subtract the amount that is being transported out from it. For convergence, this difference should of course be zero, since there are no sources inside the domain. In mathematical terms, the total flux (convection plus diffusion) is given as F and this criterion is expressed as

$$|F_W + F_E| < \varepsilon_{\text{thres}}.$$

It must also be mentioned here that $\varepsilon_{\text{thres}}$ can take different numerical values for different variables. The numerical values of $\varepsilon_{\text{thres}}$ are often in the range 10^{-3}–10^{-6} for single precision modes depending on which definition of a convergent solution is used. Probably, the best way to ensure convergence is a combination of the criteria described above. Being able to determine convergence is one of the most crucial parts of CFD. Almost all commercial codes have various built-in functions to check convergence, but nevertheless it is often worth the extra effort to do some extra manual checks. Remember that a solution that has not converged is an 'incorrect' solution.

More about convergence can be found in Chapter 7.

Question

(1) In Examples 1 and 2 the same problem was solved in two completely different ways. Which solution procedure puts the highest demand on the computer in terms of memory usage and processor performance?

3.8 Discretization schemes

Look back at Example 1. When Eq. (3.18) was solved for the cells, the numerical values of a_E, a_W and a_P were used. These were obtained from Eq. (3.19). In turn, these expressions relied on the assumptions made in Eqs. (3.13) and (3.14). A linear interpolation was assumed in order to get accurate face values for ϕ. On the other hand, it was later argued that this was an assumption that lacks physical reliability in cases with strong convection. This means that the western-face value of, for example, cell 4 should be more influenced by cell 3 than by cell 4 as a consequence of the flow direction.

To investigate the possible effects of non-physical assumptions, Example 1 is solved yet again, now with an increased flow rate.

3.8.1 Example 3 – increased velocity

The same data as in Example 1 are used, but the gas velocity is now 5 cm s^{-1}. Using the same solution procedure as in Example 2, Eq. (3.19) gives

$$\begin{aligned} a_E &= -0.0150, \\ a_W &= 0.0350, \\ a_P &= 0.0200. \end{aligned} \quad (3.21)$$

Equation (3.18) is now solved for all cells. One Gauss–Seidel iteration with the same starting guess as in Example 2 gives $\phi_1 = -90.0000$, $\phi_2 = -195.0000$, $\phi_3 = -378.7500$, $\phi_4 = -700.3125$, $\phi_5 = -1263.0469$, $\phi_6 = -2247.8320$, $\phi_7 = -3971.2061$, $\phi_8 = -6987.1106$, $\phi_9 = -12\,264.9435$ and $\phi_{10} = -21\,576.1512$.

If some more iterations are run, the numerical values of ϕ will become even larger. The values will fluctuate between very large positive values and very large negative values from iteration to iteration. This behaviour is typical for a *diverged* solution. A diverged solution gives incorrect results. In this case it is quite obvious since ϕ represents the molar concentration of species A, and of course can't take negative values. The question one should ask is that of why the solution diverged. The only difference from the successfully solved Example 2 is that the flow rate was increased and thereby the convective transport became more dominant.

Perhaps a look at the value of a_E, Eq. (3.21), can give a hint regarding what's wrong. It has already been concluded that the value of the coefficient is a measure of the strength of the interaction between the eastern and present cells. A high number means a strong interaction, and vice versa. With no interaction, the coefficient should be zero. A value below zero lacks physical reliability. Thus, there is reason to suspect that the failure stems from the 'incorrect' coefficient.

Equation (3.19) has been used to calculate the coefficients. These relations build on the assumption that the face values of ϕ can be estimated using a mean value of the neighbouring cells. As shown previously, this assumption cannot be used in cases with strong convection. In the present case, strong convection means that

$$a_E < 0 \rightarrow \frac{\rho U}{2} > \frac{\Gamma}{x_E - x_P}. \quad (3.22)$$

This criterion can also be expressed in terms of a dimensionless number. The Péclet number is defined as the ratio between convective mass transfer and diffusive mass transfer,

$$Pe = \frac{\rho U}{\Gamma/(x_E - x_P)} = \frac{\rho U(x_E - x_P)}{\Gamma}. \quad (3.23)$$

The criterion for strong convection would then be (cf. Eq. (3.22))

$$|Pe| > 2. \tag{3.24}$$

Here, it is assumed that the velocity is always positive, i.e. the flow is from west to east. However, the flow could just as well go in the opposite direction, giving a negative value of a_W instead. This would also have resulted in a divergent solution. The absolute sign for *Pe* in Eq. (3.24) evolves from this fact. In Example 3, $Pe = 5$, which satisfies Eq. (3.24) and hence makes the problem diverge.

This example illustrates that it's important to keep in mind that all assumptions must have physical reliability. This will always be important in CFD, not only in the numerical aspects, but also in other parts of the CFD problem. When the various turbulence models are introduced in following chapters it is stated that each model has its physical limitations and that overlooking these limitations can result in an incorrect solution. Bearing in mind the physical background of the problem is thus always important.

3.8.2 Boundedness and transportiveness

A desired property of a discretization scheme is that it should uphold *boundedness*. A *bounded* variable has a value that is neither larger nor smaller than any of the values that are used to calculate it. In Example 1 central differencing was used to estimate the face values of ϕ. The face value was then simply the mean value of the two surrounding cells. Thus, the face values are bounded. The cell values were calculated using Eq. (3.18), which actually is nothing other than a weighted mean value of the two neighbouring cells. This follows from the fact that $a_W + a_E = a_P$. On the other hand, Eq. (3.18) can be seen as a mean value only for cases with positive values of the coefficients. If one of the values is negative, it is no longer possible to say that ϕ_P is bounded, since it can take values that are either larger or smaller than those for any of the neighbouring cells. Thus, the central-differencing scheme is said to be *conditionally bounded*. It should be noted that there are more advanced central-differencing schemes that are bounded, but they are beyond the scope of this book.

Transportiveness is another desired property of the schemes. It has already been discussed briefly, but not defined properly. The ability of the numerical scheme to 'feel' in what direction the information is being transported is a description of transportiveness. As previously discussed, the solution in Example 3 failed partly due to the lack of this property. Transportiveness is linked to boundedness in this case, since lack of transportiveness gave an unbounded solution. The central-differencing scheme calculates the face values without taking any notice of the direction of the flow, i.e. the direction of the information. Thus, the central-differencing scheme does not satisfy the transportiveness requirement.

3.8.3 The upwind schemes

From Example 3, and the discussion above, it is clear that in cases with strong convection the face values of ϕ have to be estimated in some other way than using Eq. (3.13). A reasonable line is to suggest that the face values between, for example, cell 4 and cell 5 are dependent only upon cell values from cell 4 or cells further upstream. Schemes that let face values be dependent only on upstream conditions are called *upwind schemes*. We will now examine two of these schemes closer.

First-order upwind

The idea behind the first-order upwind scheme is to have physical reliability for convective flows simply by letting the face value between two cells be equal to the value for the nearest upstream cell, i.e.

$$\begin{aligned} \phi_{\mathrm{w}} &= \phi_{\mathrm{W}}, \\ \phi_{\mathrm{e}} &= \phi_{\mathrm{P}}, \end{aligned} \tag{3.25}$$

or, for negative velocities,

$$\begin{aligned} \phi_{\mathrm{w}} &= \phi_{\mathrm{P}}, \\ \phi_{\mathrm{e}} &= \phi_{\mathrm{E}}. \end{aligned} \tag{3.26}$$

The gradients are still estimated using Eq. (3.14).

If we return to our previous examples and use the first-order upwind scheme instead of the central-differencing scheme, we can rewrite Eq. (3.15) as

$$[(\rho U \phi_{\mathrm{P}}) - (\rho U \phi_{\mathrm{W}})] = \left[\left(\Gamma \frac{\phi_{\mathrm{E}} - \phi_{\mathrm{P}}}{x_{\mathrm{E}} - x_{\mathrm{P}}} \right) - \left(\Gamma \frac{\phi_{\mathrm{P}} - \phi_{\mathrm{W}}}{x_{\mathrm{P}} - x_{\mathrm{W}}} \right) \right]. \tag{3.27}$$

The physical meaning of the terms in the equation is the same as before; the difference in convective transport of ϕ is balanced out by the difference in diffusion. The only difference from Eq. (3.15) is that face values of ϕ have been expressed using the first-order upwind scheme instead of the central-differencing scheme. If we write Eq. (3.27) in the same form as Eq. (3.18), we get

$$\begin{aligned} a_{\mathrm{E}} &= \frac{\Gamma}{x_{\mathrm{E}} - x_{\mathrm{P}}}, \\ a_{\mathrm{W}} &= \frac{\Gamma}{x_{\mathrm{P}} - x_{\mathrm{W}}} + \rho U, \\ a_{\mathrm{P}} &= \rho U + \frac{\Gamma}{x_{\mathrm{E}} - x_{\mathrm{P}}} + \frac{\Gamma}{x_{\mathrm{P}} - x_{\mathrm{W}}}. \end{aligned} \tag{3.28}$$

Then, 40 iterations with the GSA will yield the following values for the cells: $\phi_1 = 10.0004$, $\phi_2 = 10.0003$, $\phi_3 = 10.0003$, $\phi_4 = 10.0007$, $\phi_5 = 10.0034$, $\phi_6 = 10.0199$, $\phi_7 = 10.1191$, $\phi_8 = 10.7143$, $\phi_9 = 14.2858$ and $\phi_{10} = 35.7143$.

The results are as expected; the high flow rate makes the western boundary more dominant. Almost all cells in the domain, except the most eastern, have much the same value of ϕ of approximately 10. Again, this seems reasonable.

From Eq. (3.28) it follows that the first-order upwind scheme is bounded. It also fulfils the requirement of transportiveness since care is taken regarding the direction of the flow, cf. Eqs. (3.25) and (3.26). However, it also overestimates the transport of entities in the flow direction. This gives rise to so-called *numerical diffusion*.

Second-order upwind

To improve accuracy – we will discuss accuracy in more detail later – there is an upwind scheme that predicts the face values using information from two upwind cells. To estimate the eastern-face value, the scheme assumes that the gradient between the present cell and the eastern face is the same as that between the western cell and the present cell. In mathematical terms,

$$\frac{\phi_e - \phi_P}{x_e - x_P} = \frac{\phi_P - \phi_W}{x_P - x_W} \rightarrow \phi_e = \frac{(\phi_P - \phi_W)(x_e - x_P)}{x_P - x_W} + \phi_P. \tag{3.29}$$

For an equidistant grid, Eq. (3.29) gives that

$$\phi_e = 1.5\phi_P - 0.5\phi_W. \tag{3.30}$$

A major drawback with the second-order upwind scheme is that it is *unbounded.* To avoid the numerical problems that often arise as a result of unbounded schemes, some *bounded* second-order schemes have been developed, e.g. the van Leer scheme. The definitions will be stated here.

The van Leer scheme

If $|\phi_E - 2\phi_P + \phi_W| \leq |\phi_E - \phi_W|$, the value of ϕ at the eastern face is (cf. Eq. (3.29))

$$\phi_e = \phi_P + \frac{(\phi_E - \phi_P)(\phi_P - \phi_W)}{\phi_E - \phi_W}. \tag{3.31}$$

Otherwise (cf. Eq. (3.25)),

$$\phi_e = \phi_P. \tag{3.32}$$

The velocity is assumed to be positive. The van Leer scheme implements the unbounded second-order upwind scheme, Eq. (3.31), if the gradient is 'smooth', i.e. the second derivative of ϕ is 'small'. Otherwise, the first-order upwind scheme, Eq. (3.32), is used.

3.8.4 Taylor expansions

Before proceeding, a short mathematical review of Taylor expansions will be given.

Taylor's theorem for a 1D expansion of a real function $f(x)$ about a point $x = x_0$ is given without a proof:

$$f(x) = f(x_0) + (x - x_0)f'(x_0) + \frac{(x - x_0)^2}{2!} f''(x_0) + \cdots$$
$$+ \frac{(x - x_0)^n}{n!} f^{(n)}(x_0) + \int_{x_0}^{x} \frac{(x - u)^n}{n!} f^{(n+1)}(u) \mathrm{d}u. \tag{3.33}$$

The last term in Eq. (3.33) is called the *Lagrange remainder*. Taylor's theorem, except for the Lagrange remainder, was devised by the English mathematician Brook Taylor in 1712 and published in *Methodus in crementorum directa et inversa* in 1715. The more terms are included in the series, the more accurate the estimation will be. Taylor expansions will be used when discussing accuracy.

3.8.5 Accuracy

Two different types of discretization schemes have been presented, the central-differencing scheme and the upwind schemes, one that fulfils the requirement of transportiveness and one that does not. The last section concluded that for problems with strong convection the central-differencing scheme failed. The first-order upwind scheme was then used instead. On the other hand, the first-order upwind scheme used only one cell to estimate the face value, compared with two cells for the central-differencing scheme, and thus we should expect the first-order upwind scheme to be less *accurate* than the central-differencing scheme. Generally, the more information is used, the better the estimation. Accuracy can be quantified in several ways.

For the central-differencing scheme, we have

$$\begin{aligned} \phi_\mathrm{w} &= \frac{\phi_\mathrm{W} + \phi_\mathrm{P}}{2}, \\ \phi_\mathrm{e} &= \frac{\phi_\mathrm{P} + \phi_\mathrm{E}}{2} \end{aligned} \tag{3.13}$$

and

$$\begin{aligned} \left(\frac{\mathrm{d}\phi}{\mathrm{d}x}\right)_\mathrm{w} &= \frac{\phi_\mathrm{P} - \phi_\mathrm{W}}{x_\mathrm{P} - x_\mathrm{W}}, \\ \left(\frac{\mathrm{d}\phi}{\mathrm{d}x}\right)_\mathrm{e} &= \frac{\phi_\mathrm{E} - \phi_\mathrm{P}}{x_\mathrm{E} - x_\mathrm{P}}. \end{aligned} \tag{3.14}$$

If it is assumed that the grid is equidistant (this has been the case in all our examples so far), the *grid spacing* can be defined as $\Delta x = x_\mathrm{P} - x_\mathrm{W} = x_\mathrm{E} - x_\mathrm{P}$. If a Taylor expansion of ϕ_E and ϕ_P is made about x_e, the following result is reached:

$$\begin{aligned} \phi_\mathrm{E} = \phi_\mathrm{e} &+ (\Delta x/2)\left(\frac{\mathrm{d}\phi}{\mathrm{d}x}\right)_\mathrm{e} + \frac{(\Delta x/2)^2}{2}\left(\frac{\mathrm{d}^2\phi}{\mathrm{d}x^2}\right)_\mathrm{e} + \frac{(\Delta x/2)^3}{6}\left(\frac{\mathrm{d}^3\phi}{\mathrm{d}x^3}\right)_\mathrm{e} \\ &+ \frac{(\Delta x/2)^4}{24}\left(\frac{\mathrm{d}^4\phi}{\mathrm{d}x^4}\right)_\mathrm{e} + O\left[(\Delta x)^5\right], \end{aligned} \tag{3.34}$$

$$\begin{aligned} \phi_\mathrm{P} = \phi_\mathrm{e} &- (\Delta x/2)\left(\frac{\mathrm{d}\phi}{\mathrm{d}x}\right)_\mathrm{e} + \frac{(\Delta x/2)^2}{2}\left(\frac{\mathrm{d}^2\phi}{\mathrm{d}x^2}\right)_\mathrm{e} - \frac{(\Delta x/2)^3}{6}\left(\frac{\mathrm{d}^3\phi}{\mathrm{d}x^3}\right)_\mathrm{e} \\ &+ \frac{(\Delta x/2)^4}{24}\left(\frac{\mathrm{d}^4\phi}{\mathrm{d}x^4}\right)_\mathrm{e} + O\left[(\Delta x)^5\right]. \end{aligned} \tag{3.35}$$

Here, $O[(\Delta x)^n]$ is the *truncation error*. Next, Eqs. (3.34) and (3.35) are inserted into the right-hand side of the second equation in Eqs. (3.13) and (3.14), which results in

$$\frac{\phi_\mathrm{P} + \phi_\mathrm{E}}{2} = \phi_\mathrm{e} + \frac{(\Delta x)^2}{8}\left(\frac{\mathrm{d}^2\phi}{\mathrm{d}x^2}\right)_\mathrm{e} + \frac{(\Delta x)^4}{384}\left(\frac{\mathrm{d}^4\phi}{\mathrm{d}x^4}\right) + O\left[(\Delta x)^6\right], \tag{3.36}$$

$$\frac{\phi_\mathrm{E} - \phi_\mathrm{P}}{\Delta x} = \left(\frac{\mathrm{d}\phi}{\mathrm{d}x}\right)_\mathrm{e} + \frac{(\Delta x)^2}{24}\left(\frac{\mathrm{d}^3\phi}{\mathrm{d}x^3}\right)_\mathrm{e} + O\left[(\Delta x)^5\right]. \tag{3.37}$$

According to Eqs. (3.13) and (3.14), these expressions should be equal to the face value of ϕ and the gradient of ϕ at the eastern face, respectively, *assuming that central differencing is used*. Thus, since the second-order derivative $d^2\phi/dx^2$ is unknown,

$$\phi_e = \phi_e^{CD} + O\left[(\Delta x)^2\right] \tag{3.38}$$

and

$$\left(\frac{d\phi}{dx}\right)_e = \left(\frac{d\phi}{dx}\right)_e^{CD} + O\left[(\Delta x)^2\right]. \tag{3.39}$$

Before we comment on the results, we repeat the same procedure for the first-order upwind scheme. Since the face gradient is predicted in the same way as with central differencing, the face-value estimation, i.e. Eq. (3.25), must be examined. According to this relation, the face value of the eastern face is simply equal to the cell value in the present cell. A Taylor expansion of ϕ about x_P gives

$$\phi_e = \phi_P + (\Delta x/2)\left(\frac{d\phi}{dx}\right)_P + \frac{(\Delta x/2)^2}{2}\left(\frac{d^2\phi}{dx^2}\right)_P + O\left[(\Delta x)^3\right]. \tag{3.40}$$

Here, the first-order derivative is unknown and the outcome is

$$\phi_e = \phi_e^{1u} + O(\Delta x). \tag{3.41}$$

If Eqs. (3.38) and (3.41) are compared, it can be seen that, for a reduction of Δx, the face-value estimation seems to approach the 'true' value quicker in the case with the central-differencing scheme. In other words, this means that, if the grid is made denser, i.e. more cells are introduced, the error in the central-differencing scheme will be reduced more quickly than in the case with the first-order upwind scheme. The lowest order of the grid spacing in the central-differencing scheme is 2, hence the central-differencing scheme is referred to as *second-order accurate*. For the first-order upwind scheme, the corresponding number is 1; hence this scheme is referred to as *first-order accurate*.

Higher-order schemes are more accurate but this can be at the expense of numerical stability. When stability is problematic it is recommended that one start with a simple first-order upwind scheme and change to a higher-order scheme after some iterations. Going to higher order is always necessary in order to minimize numerical diffusion when the grids are not aligned with the flow.

3.8.6 The hybrid scheme

An attempt to combine the positive properties of both the central-differencing scheme and the first-order upwind scheme has been proposed. This scheme is called the *hybrid differencing scheme*. This scheme implements the upwind scheme at faces where the criterion in Eq. (3.24) is fulfilled, and uses central differencing elsewhere. Thus, this scheme takes advantage of both the high accuracy of the central-differencing scheme and the more physical properties in terms of boundedness and transportiveness of the upwind scheme.

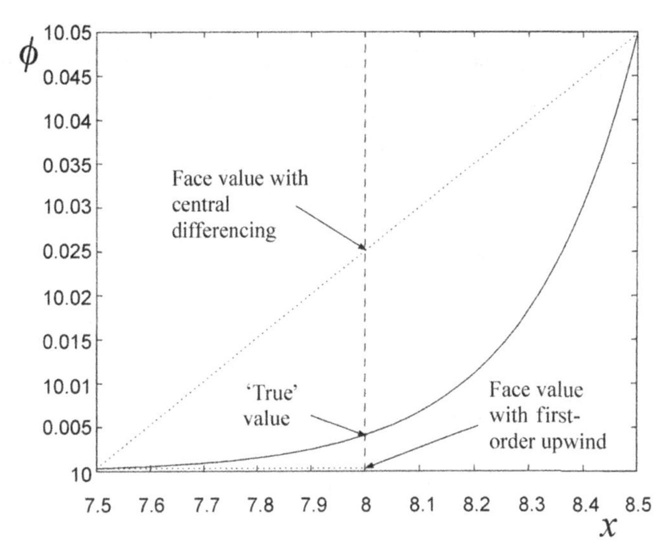

Figure 3.10 Face-value approximations of the face between cells 8 and 9 with central differencing and first-order upwind, respectively. The analytical ('true') solution which, in this specific case, corresponds to the power-law prediction has been added for comparison. $U = 5$ cm s^{-1}.

3.8.7 The power-law scheme

A more accurate scheme is the *power-law scheme*. Briefly, the face value of ϕ is estimated by solving a convection–diffusion equation (cf. Eq. (3.1)),

$$\rho \frac{\mathrm{d}(U\phi)}{\mathrm{d}x} = \frac{\mathrm{d}}{\mathrm{d}x}\left(\Gamma \frac{\mathrm{d}\phi}{\mathrm{d}x}\right). \tag{3.42}$$

Equation (3.42) has the following solution, assuming constant fluid and flow properties:

$$\frac{\phi(x) - \phi_0}{\phi_{\Delta x} - \phi_0} = \frac{\exp(\rho U x / \Gamma) - 1}{\exp(\rho U \, \Delta x / \Gamma) - 1}, \tag{3.43}$$

where the indices 0 and Δx represent two neighbouring cells. If the grid spacing is equidistant, the face is situated at $x = 0.5$. The face value can now be estimated using Eq. (3.43). If the parameters from the previous examples are used, the plot of $\phi(x)$ against x will clearly show that for $U = 5$ cm s^{-1} the face value is almost exactly the same as for the upstream cell; see Figure 3.10. The central-differencing scheme makes a bad estimation of the face value. This was the reason why the upwind scheme was chosen in Example 3.

3.8.8 The QUICK scheme

Numerous successful attempts have been made to create numerical schemes with higher accuracy than second order. One of them will very briefly be discussed, namely QUICK (Quadratic Upstream Interpolation for Convective Kinetics). QUICK combines the strengths of both the upwind schemes and of the central differencing; it uses three-point upstream quadratic interpolation to estimate the face values. For the western and eastern faces, respectively,

$$\phi_{\mathrm{w}} = \frac{6}{8}\phi_{\mathrm{W}} + \frac{3}{8}\phi_{\mathrm{P}} - \frac{1}{8}\phi_{\mathrm{WW}}, \tag{3.44}$$

$$\phi_{\mathrm{e}} = \frac{6}{8}\phi_{\mathrm{P}} + \frac{3}{8}\phi_{\mathrm{E}} - \frac{1}{8}\phi_{\mathrm{W}}. \tag{3.45}$$

It has been assumed here that the velocity is positive. A Taylor expansion of ϕ around the eastern face gives

$$\phi_{\mathrm{E}} = \phi_{\mathrm{e}} + (\Delta x/2)\left(\frac{\mathrm{d}\phi}{\mathrm{d}x}\right)_{\mathrm{e}} + \frac{(\Delta x/2)^2}{2}\left(\frac{\mathrm{d}^2\phi}{\mathrm{d}x^2}\right)_{\mathrm{e}} + O\left[(\Delta x)^3\right], \tag{3.34$'$}$$

$$\phi_{\mathrm{P}} = \phi_{\mathrm{e}} - (\Delta x/2)\left(\frac{\mathrm{d}\phi}{\mathrm{d}x}\right)_{\mathrm{e}} + \frac{(\Delta x/2)^2}{2}\left(\frac{\mathrm{d}^2\phi}{\mathrm{d}x^2}\right)_{\mathrm{e}} + O\left[(\Delta x)^3\right], \tag{3.35$'$}$$

$$\phi_{\mathrm{W}} = \phi_{\mathrm{e}} - (1.5\Delta x)\left(\frac{\mathrm{d}\phi}{\mathrm{d}x}\right)_{\mathrm{e}} + \frac{(1.5\Delta x)^2}{2}\left(\frac{\mathrm{d}^2\phi}{\mathrm{d}x^2}\right)_{\mathrm{e}} + O\left[(\Delta x)^3\right]. \tag{3.46}$$

On inserting Eqs. (3.34′), (3.35′) and (3.46) into Eq. (3.45), we get

$$\phi_{\mathrm{e}} = \phi_{\mathrm{e}}^{\mathrm{QUICK}} + O\left[(\Delta x)^3\right]. \tag{3.47}$$

Thus, the QUICK scheme is *third-order accurate*. Further, it can be shown that it is *unbounded* but fulfils the *transportiveness* criterion. The use of QUICK is restricted to hexahedral meshes.

3.8.9 More advanced discretization schemes

The discretization schemes mentioned previously are only a few among the many available. In most commercial CFD codes there are many variations of schemes, and it is not our intention to provide a complete list in Table 3.1. However, two more will be mentioned here without going into too much detail.

- MUSCL (Monotone Upstream-centred Schemes for Conservation Laws). The MUSCL scheme shows a similar degree of accuracy to QUICK, but is not limited to hexahedral meshes.
- HRIC (High-Resolution Interface Capturing). The HRIC scheme is primarily used in multiphase flows when tracking the interface between the phases. HRIC has proven to be more accurate than QUICK for VOF simulations (see Chapter 6).

For all of the discretization schemes that have been presented here, the error tends to zero as the grid spacing is reduced infinitely. Schemes that uphold this property are said to be *consistent*. It should also be mentioned that the definitions of some of the discretization schemes presented here seem to differ in the literature.

Table 3.1 Discretization schemes

Discretization scheme	Advantages and disadvantages
Central	Works well when diffusion dominates. Bounded variants are recommended for LES simulations.
First-order upwind	Good when convection dominates and the flow is aligned with the grid. Bounded and robust, and a good scheme to start off the calculation. May introduce numerical diffusion and should be replaced with higher-order schemes for the final calculations.
Second-order upwind	Good for all Péclet numbers, but it is unbounded and not as robust as first-order upwind.
Power law	Good for intermediate values of the Péclet number ($Pe < 10$).
QUICK	Good for all Péclet numbers. Better accuracy than the second-order scheme for rotating or swirling flows. In general the second-order scheme is sufficient. Applicable only to quadrilateral or hexahedral meshes.
MUSCL	Better accuracy than the second-order scheme for rotating or swirling flows. The scheme is used on all types of meshes.
HRIC	Used primarily for interface tracking in VOF simulations. Better accuracy than QUICK.

3.9 Solving the velocity field

In the previous examples and discussions it has been assumed that the velocity field has somehow been determined; see Eq. (3.19) as an illustrative example of this. Example 1 also concluded that the predetermined velocity gave rise to a set of *linear* equations. However, most commonly the user does not know the velocity from the start, and hence has to solve for it as well. Solving the velocity field requires some extra attention; we will examine this now.

The transport equations for momentum (also called Navier–Stokes equations) and the continuity equation are

$$\frac{\partial \rho U_i}{\partial t} + \frac{\partial \rho U_i U_j}{\partial x_j} = \frac{\partial}{\partial x_j}\left(\mu \frac{\partial U_i}{\partial x_j}\right) - \frac{\partial P}{\partial x_i} \tag{3.48}$$

and

$$\frac{\partial \rho}{\partial t} + \frac{\partial \rho U_j}{\partial x_j} = 0. \tag{3.49}$$

These expressions were derived in Chapter 2. As can be seen, the momentum equation is very similar to the transport equations that have been solved in previous examples. There is a small difference, though, that makes the solution procedure much more complicated than in previous cases, and that is the fact that there now exists a source term in the equation. The source term has the formulation of a pressure gradient, which shouldn't be surprising. A pressure gradient in e.g. a tube gives rise to the momentum in the streamwise direction. Summing the number of variables in Eqs. (3.48) and (3.49), there are four variables, namely three velocities and one pressure (remember that we have assumed incompressible flow!). The total number of equations is four (3 + 1), so it should be possible to solve the system straight away. Normally, Eq. (3.48) is used for the three velocity components. Unfortunately, Eq. (3.49) cannot be used to solve for pressure. Hence, the equation must be modified first. Taking the divergence of Eq. (3.48) gives

$$\frac{\partial^2}{\partial x_i\,\partial t}(\rho U_i) + \frac{\partial^2}{\partial x_i\,\partial x_j}\left(\rho U_i U_j\right) = \frac{\partial^2}{\partial x_i\,\partial x_j}\left(\mu \frac{\partial U_i}{\partial x_j}\right) - \frac{\partial^2 P}{\partial x_i\,\partial x_i}. \tag{3.50}$$

The first term on the left-hand side and the first term on the right-hand side can be rewritten (assuming constant density and viscosity) as

$$\frac{\partial}{\partial x_i}\left(\frac{\partial \rho U_i}{\partial t}\right) = \frac{\partial}{\partial t}\left(\frac{\partial \rho U_i}{\partial x_i}\right),$$

$$\frac{\partial}{\partial x_i}\left[\frac{\partial}{\partial x_j}\left(\mu \frac{\partial U_i}{\partial x_j}\right)\right] = \frac{\partial}{\partial x_j}\left[\frac{\partial}{\partial x_j}\left(\mu \frac{\partial U_i}{\partial x_i}\right)\right].$$

Owing to continuity (see Eq. (3.49)), the right-hand sides of both these expressions are equal to zero. Equation (3.50) can thus be written in the following way:

$$\frac{\partial^2 P}{\partial x_i\,\partial x_i} = -\frac{\partial^2}{\partial x_i\,\partial x_j}\left(\rho U_i U_j\right). \tag{3.51}$$

Equation (3.51), which is a scalar equation, can be used as a direct equation for pressure. However, numerical problems are commonly encountered, for reasons that will not be discussed in this book. Therefore, in almost all commercial CFD software an iterative procedure for solving Eq. (3.51) is adopted, thereby avoiding the numerical problems. Some of the most widely used algorithms are SIMPLE, SIMPLEC, SIMPLER and PISO. The algorithms of the SIMPLE (Semi-Implicit Method for Pressure-Linked Equations) family use a starting guess for pressure and velocity to solve for the corresponding velocities via the momentum equations. Since the starting guess of the pressure will not be correct, the velocities obtained will not satisfy continuity. Hence, correction factors for pressure and velocity are introduced, and transport equations for these factors are proposed and solved to give corrected velocities and pressure. The other transport equations, e.g. for various scalars, are then solved. Then, a check for convergence is made and the procedure is repeated until convergence is reached. SIMPLER and SIMPLEC are improved variants of SIMPLE. The PISO algorithm is mostly used for unsteady flow. It can be seen as an extension of the SIMPLE algorithm, but it uses two levels of correction instead of one as is the case with SIMPLE.

It is generally not possible to say that a specific scheme is always better than another with respect to efficiency or robustness. These properties are very dependent on the flow conditions at hand, but a proper choice of scheme can sometimes speed up the simulations significantly. The performance and behaviour of the schemes have been thoroughly examined, and a few conclusions are mentioned here.

For a laminar backward-facing step it has been shown that PISO performs twice as fast as SIMPLE, while it has been shown to be four times as slow for a case concerning flow through a heated fin. For steady flow problems, SIMPLER has been shown to need

30%–50% less computer time than SIMPLE to solve problems in general. The degree of coupling between the momentum equations and scalar equations has been shown to have a significant impact sometimes. In problems where there is only a weak coupling or no coupling at all between the momentum and scalar equations, PISO has been shown to have the most robust convergence and is quicker than SIMPLEC and SIMPLER. If the opposite is valid, i.e. there is a strong coupling between the momentum and scalar equations, SIMPLER and SIMPLEC have been shown to perform better than PISO. Further, it has not been possible to claim that either SIMPLER or SIMPLEC is superior to the other in general. SIMPLEC usually performs better in situations in which the rate of convergence is limited by the pressure–velocity coupling, such as non-complex laminar-flow cases.

In order to solve the transport equations for momentum, Eq. (2.21), it is required that the pressures at the cell faces are known. As in the cases with other transported quantities, this is achieved by interpolation of the values in the neighbouring cells. Most commercial CFD software offers a variety of interpolation schemes. Each scheme uses its own interpolation function.

The standard scheme uses the coefficients a_{p}, a_{e}, etc. to interpolate the pressure on the cell faces. As long as the pressure variation between the cells is smooth, the scheme usually works fine. For flows with large body forces the standard scheme has been shown to be troublesome. In such cases it is recommended that the body force-weighted scheme be used. This scheme assumes that the normal gradient of the difference between the body force and pressure is constant. For swirling flows and flows with natural convection, the PRESTO! scheme should be used. The PRESTO! scheme uses the discrete continuity equation to calculate the pressure field on a mesh that is geometrically shifted so that the new cell centres are where the faces of the ordinary mesh are placed. This means that the pressures on the faces are now known. The PRESTO! scheme is also recommended for VOF simulations. The second-order pressure scheme is analogous to the second-order discretization scheme presented earlier and is recommended for compressible flows. The second-order scheme can be numerically unstable if it is used at the start of a calculation or on a bad mesh.

3.9.1 Under-relaxation

The momentum equations, Eq. (3.48), contain nonlinear terms, e.g. ρV^2 in the momentum equation for V. To avoid divergence due to nonlinearities, so-called *under-relaxation* is used in the solving procedure. The under-relaxation factor α is defined as

$$\phi_{\mathrm{new}} = \alpha\phi_{\mathrm{solver}} + (1-\alpha)\,\phi_{\mathrm{old}}, \tag{3.52}$$

where ϕ_{new} is the new value at the cell, ϕ_{solver} is the value for the solution to the last iteration, and ϕ_{old} is the previous value at the cell that is used to compute ϕ_{solver}. Thus, a large value of α means that the new value will be very influenced by the solved value, and vice versa. α lies usually in the range between zero and one. In many cases the under-relaxation factor can be increased during iterations. Each transport equation or pressure equation has its 'own' corresponding under-relaxation factor. Too low a value

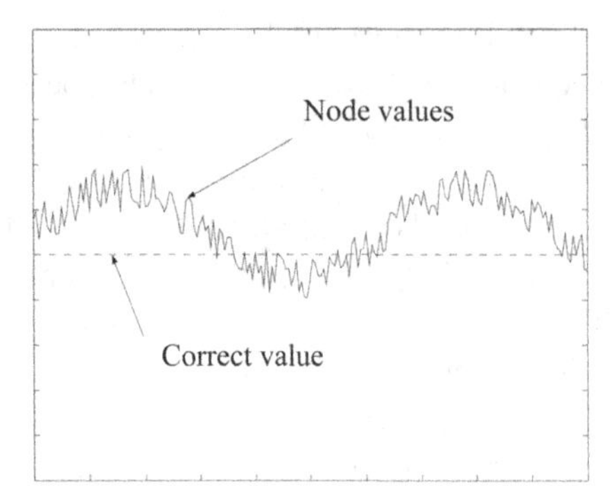

Figure 3.11 High- and low-frequency errors. The correct values correspond to a constant value throughout the domain.

of the under-relaxation factor gives unmotivated long computational times. Too large a value can give a diverged solution. Thus, choosing optimal values is of great practical importance.

Questions

(1) Since the central-differencing scheme uses a weighted average (see Eq. (3.18)) of the neighbouring cells to calculate the present cell, by definition it cannot be unbounded. This is obviously wrong! What restrictions must be put on $a_{\mathrm{E}}/a_{\mathrm{P}}$ and $a_{\mathrm{W}}/a_{\mathrm{P}}$ in order to keep ϕ_{P} bounded, i.e. between ϕ_{E} and ϕ_{W}.

(2) Describe what factors determining what discretization scheme should be used in a specific situation. Can these factors always be determined before the simulation has been performed?

(3) Is a higher-order differencing scheme always better than a scheme with lower order? Consider robustness, accuracy, CPU time and memory demand.

3.10 Multigrid

An efficient way to enhance convergence speed is to use a so-called multigrid solver. The idea is to use at least two levels of grid spacing. Take the previous examples as illustration! Since the Gauss–Seidel solver calculates the cell value only by using the values of the neighbouring cells, it is very efficient in eliminating local errors, see also Figure 3.11. However, there could be situations in which there is an extra need for information to be transported fast through the domain. Let's say, for instance, that the pressure is increased in an area within the domain. This pressure increase will affect the entire domain instantly. In the CFD simulation the pressure increase will be transported as the Gauss–Seidel algorithm sweeps through the domain. However, if the computational domain contains many cells, it will take a significant time before

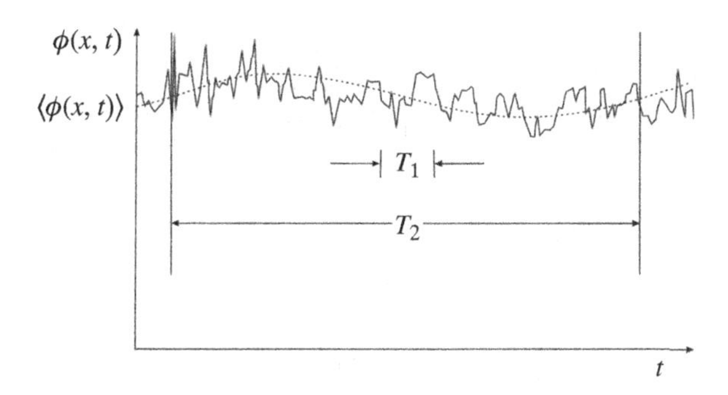

Figure 3.12 The mean and fluctuation of an instantaneous flow variable as a function of time. If $T_1 \ll T_2$, the mean flow and fluctuating quantities are uncorrelated.

information is transported from cells at one end to cells at the other. This will mean that much iteration has to be performed before a converged solution is obtained. Here is where the different grid levels come in. By using a level with larger cells, information can be transported faster than would be possible without this level. Information is transferred from coarser to denser grid levels via so-called *prolongation* and in the other direction by the so-called *restriction process*. Altogether, this means that the rate of convergence is enhanced, often by as much as several orders of magnitude.

3.11 Unsteady flows

Before starting to discuss unsteady flows, we need clarity regarding the definition of the term 'unsteady'. Defining an unsteady flow as a flow that shows variation in time would imply that almost all flows are unsteady. Remember that most flows of industrial importance are turbulent and that turbulence is always time-dependent! Hence, the definition of unsteadiness must be tuned. In most CFD problems, the flow studied is turbulent, forcing one to solve for mean quantities instead of instantaneous ones and to use an appropriate turbulence model to model the interaction between sub-scale motions and the mean flow (see Chapter 4). In these cases unsteadiness must mean that the mean quantities rather than the instantaneous quantities change with time. Further, it is required that there is a *large* separation in timescales to be able to distinguish between turbulent fluctuations and mean flow fluctuations, see also Figure 3.12. Unfortunately, in most engineering applications concerning unsteady flows this criterion is not completely fulfilled, giving a non-zero correlation between the mean flow and the fluctuating quantity. In such cases, turbulence is moved from the turbulence model to the unsteady term in the time-averaged Navier–Stokes equation.

The procedure for solving unsteady problems follows the same path as for steady problems. First, the general transport equation is integrated, with respect to both space and time,

$$\int_{\text{c.v.}} \left(\int_{t}^{t+\Delta t} \rho \frac{\partial \phi}{\partial t} \, \mathrm{d}t \right) \mathrm{d}V + \int_{t}^{t+\Delta t} \left[\int_{\text{c.v.}} \rho \frac{\partial (U_j \phi)}{\partial x_j} \, \mathrm{d}V \right] \mathrm{d}t$$
$$= \int_{t}^{t+\Delta t} \left[\int_{\text{c.v.}} \frac{\partial}{\partial x_j} \left(\Gamma \frac{\partial \phi}{\partial x_j} \right) \mathrm{d}V \right] \mathrm{d}t + \int_{t}^{t+\Delta t} \left(\int_{\text{c.v.}} S_\phi \, \mathrm{d}V \right) \mathrm{d}t. \quad (3.53)$$

The order of integration is reversed in the accumulation term. To clarify the steps needed, an example is given.

3.11.1 Example 4 – time-dependent simulation

Using the same experimental set-up as in Example 1, but this time without any convection, the unsteady case is solved. Initially there is no species A in the domain. The boundary conditions are the same as in the original example.

The integrated transport equation for ϕ in the present example is

$$\int_{\text{w}}^{\text{e}} \left(\int_{t}^{t+\Delta t} \rho \frac{\mathrm{d}\phi}{\mathrm{d}t} \, \mathrm{d}t \right) \mathrm{d}x = \int_{t}^{t+\Delta t} \left[\int_{\text{w}}^{\text{e}} \frac{\mathrm{d}}{\mathrm{d}x} \left(\Gamma \frac{\mathrm{d}\phi}{\mathrm{d}x} \right) \mathrm{d}x \right] \mathrm{d}t. \quad (3.54)$$

Using central differencing allows us to rewrite Eq. (3.54) as

$$\int_{\text{w}}^{\text{e}} \left(\int_{t}^{t+\Delta t} \rho \frac{\mathrm{d}\phi}{\mathrm{d}t} \, \mathrm{d}t \right) \mathrm{d}x = \int_{t}^{t+\Delta t} \left(\Gamma_{\text{e}} \frac{\phi_{\text{E}} - \phi_{\text{P}}}{\Delta x} - \Gamma_{\text{w}} \frac{\phi_{\text{P}} - \phi_{\text{W}}}{\Delta x} \right) \mathrm{d}t. \quad (3.55)$$

To proceed with the time integration an appropriate temporal discretization scheme is needed. As in the case with spatial discretization there is a choice of scheme; either 'earlier' or 'later' times (or a mixture of them) can be used to estimate the temporal derivative at the 'present' time. For now,

$$\rho[\phi_{\text{P}}(t + \Delta t) - \phi_{\text{P}}(t)]\Delta x = \int_{t}^{t+\Delta t} \left(\Gamma_{\text{e}} \frac{\phi_{\text{E}} - \phi_{\text{P}}}{\Delta x} - \Gamma_{\text{w}} \frac{\phi_{\text{P}} - \phi_{\text{W}}}{\Delta x} \right) \mathrm{d}t. \quad (3.56)$$

To continue, ϕ on the right-hand side of Eq. (3.56) is evaluated at each time step. ϕ changes in time, so there are at least three different ways to get values for ϕ: (1) let ϕ be equal to $\phi(t)$, (2) let ϕ be equal to $\phi(t + \Delta t)$, and (3) let ϕ be a mixture of the two.

We now define the weight factor θ:

$$\int_{t}^{t+\Delta t} \phi \, \mathrm{d}t = [\theta \phi(t + \Delta t) + (1 - \theta)\phi(t)]\Delta t. \quad (3.57)$$

The case $\theta = 0$ means that only the old value of ϕ is used when evaluating Eq. (3.56). This leads to so-called *explicit discretization* of this equation. The other extreme, $\theta = 1$, corresponds to a discretization where only the new value of ϕ is used in the discretization, the so-called *(fully) implicit discretization*. Of course, θ can take values also between zero and one; e.g. the *Crank–Nicolson scheme* for $\theta = 0.5$.

Using Eq. (3.57) to express ϕ_i in Eq. (3.56),

$$\rho\left[\phi_{\mathrm{P}} - \phi_{\mathrm{P}}^{\mathrm{o}}\right]\Delta x = \left[\theta\left(\Gamma_{\mathrm{e}}\frac{\phi_{\mathrm{E}} - \phi_{\mathrm{P}}}{\Delta x} - \Gamma_{\mathrm{w}}\frac{\phi_{\mathrm{P}} - \phi_{\mathrm{W}}}{\Delta x}\right)\right.$$
$$\left. + (1-\theta)\left(\Gamma_{\mathrm{e}}\frac{\phi_{\mathrm{E}}^{\mathrm{o}} - \phi_{\mathrm{P}}^{\mathrm{o}}}{\Delta x} - \Gamma_{\mathrm{w}}\frac{\phi_{\mathrm{P}}^{\mathrm{o}} - \phi_{\mathrm{W}}^{\mathrm{o}}}{\Delta x}\right)\right]\Delta t. \tag{3.58}$$

Here, ϕ_i^{o} means the value for cell i at time t. ϕ_i means the value for cell i at time $t + \Delta t$.

To examine the properties of Eq. (3.58), it is rewritten as

$$\left[\rho\frac{\Delta x}{\Delta t} + \theta\left(\frac{\Gamma_{\mathrm{e}}}{\Delta x} + \frac{\Gamma_{\mathrm{w}}}{\Delta x}\right)\right]\phi_{\mathrm{P}}$$
$$= \frac{\Gamma_{\mathrm{e}}}{\Delta x}\left[\theta\phi_{\mathrm{E}} + (1-\theta)\phi_{\mathrm{E}}^{\mathrm{o}}\right] + \frac{\Gamma_{\mathrm{w}}}{\Delta x}\left[\theta\phi_{\mathrm{W}} + (1-\theta)\phi_{\mathrm{W}}^{\mathrm{o}}\right]$$
$$+ \left[\rho\frac{\Delta x}{\Delta t} - (1-\theta)\frac{\Gamma_{\mathrm{e}}}{\Delta x} - (1-\theta)\frac{\Gamma_{\mathrm{w}}}{\Delta x}\right]\phi_{\mathrm{P}}^{\mathrm{o}}. \tag{3.59}$$

This leads to (cf. Eq. (3.19))

$$a_{\mathrm{P}}\phi_{\mathrm{P}} = a_{\mathrm{W}}\left[\theta\phi_{\mathrm{W}} + (1-\theta)\phi_{\mathrm{W}}^{\mathrm{o}}\right] + a_{\mathrm{E}}\left[\theta\phi_{\mathrm{E}} + (1-\theta)\phi_{\mathrm{E}}^{\mathrm{o}}\right]$$
$$+ \left[a_{\mathrm{P}}^{\mathrm{o}} - (1-\theta)a_{\mathrm{W}} - (1-\theta)a_{\mathrm{E}}\right]\phi_{\mathrm{P}}^{\mathrm{o}}, \tag{3.60}$$

where

$$\begin{aligned} a_{\mathrm{P}} &= \theta(a_{\mathrm{W}} + a_{\mathrm{E}}) + a_{\mathrm{P}}^{\mathrm{o}}, \\ a_{\mathrm{P}}^{\mathrm{o}} &= \rho\frac{\Delta x}{\Delta t}, \\ a_{\mathrm{W}} &= \frac{\Gamma_{\mathrm{w}}}{\Delta x}, \\ a_{\mathrm{E}} &= \frac{\Gamma_{\mathrm{e}}}{\Delta x}. \end{aligned} \tag{3.61}$$

An unsteady problem must be solved in a different way from a steady problem. This is as a consequence of having to solve the set of equations for many different times. Evidently, Eq. (3.60) must be satisfied within each time step, thus it is necessary to sub-iterate within each time step. When a convergent solution has been obtained, move forward one time step and repeat the sub-iterations. This is continued until the appropriate time has been reached. The starting guess for the next time step is the solution to the previous time step. In that way, the number of sub-iterations required within each time step usually decreases with time, provided that the time step is sufficiently small. The presence of a starting solution instead of a starting guess is required for the initial time step.

The example is finally solved using the different discretization methods.

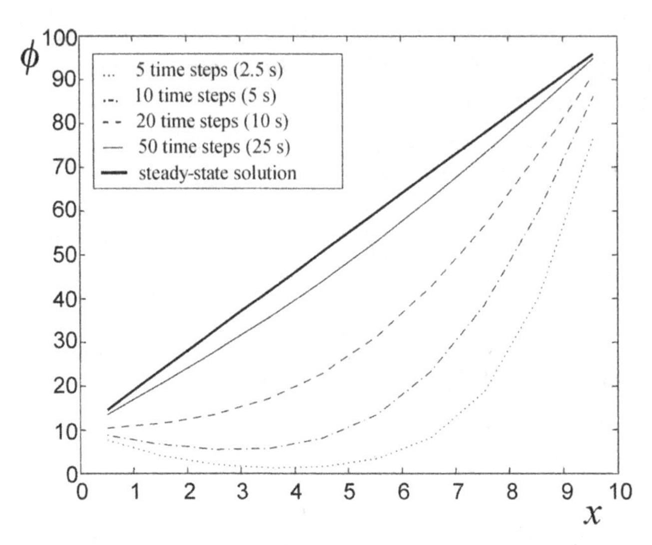

Figure 3.13 A plot of molar concentration as a function of the distance at different times. The time step is 0.4 s and the discretization method is explicit. The initial condition was $c = 0$ everywhere.

Explicit method

With $\theta = 0$, Eq. (3.60) gives

$$\phi_P = \frac{a_W \phi_W^o + a_E \phi_E^o + \left(a_P^o - a_W - a_E\right) \phi_P^o}{a_P} \tag{3.62}$$

and the coefficients

$$\begin{aligned} a_P &= a_P^o, \\ a_P^o &= \rho \frac{\Delta x}{\Delta t}, \\ a_W &= \frac{\Gamma_w}{\Delta x}, \\ a_E &= \frac{\Gamma_e}{\Delta x}. \end{aligned} \tag{3.63}$$

The next question is what time step to choose. A time step is chosen in order to be able to evaluate ϕ_P. As earlier, the scheme needs to be bounded, and therefore the time step should be chosen so that $a_P^o - a_W - a_E > 0$. This implies

$$\Delta t < \frac{\rho (\Delta x)^2}{\Gamma_e + \Gamma_w}. \tag{3.64}$$

In the example, this means that the time step cannot be larger than 0.5 seconds. Solving Eq. (3.62) with the coefficients from Eq. (3.63) using Δt from Eq. (3.64) gives a plot of the molar concentration at different times, see Figure 3.13.

The results look reasonable. Initially there is no species A in the system, and the more time elapses, the more of the species has diffused from the walls. The explicit method will be discussed more later on.

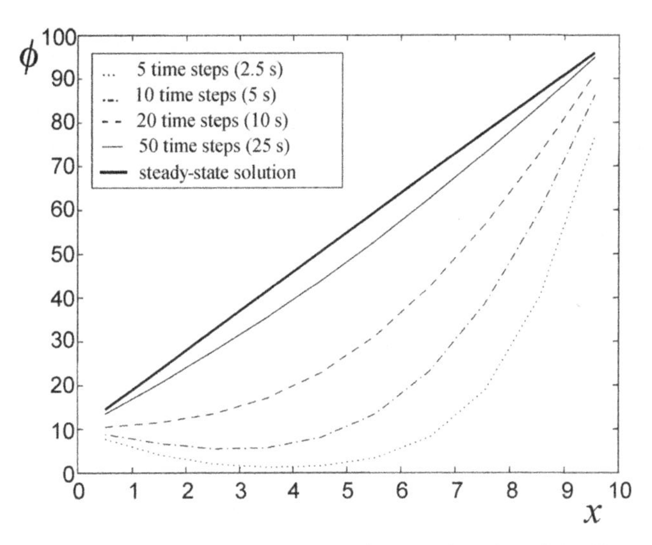

Figure 3.14 A plot of molar concentration as a function of the distance at different times. The time step is 0.5 s and the discretization method is fully implicit. The initial condition was $c = 0$ everywhere.

An advantage with the explicit method is that there is no need for sub-iterations. Looking at Eq. (3.62) explains why; all values are taken from 'old' time steps. Thus, its name *explicit*. The method is also called *global time stepping*.

Fully implicit method

As defined before, in the fully implicit method the new values of ϕ are used as an estimate of ϕ during the whole step of integration, i.e. $\theta = 1$ in this method. Equation (3.60) is then reduced to

$$a_P\phi_P = a_W\phi_W + a_E\phi_E + a_P^o\phi_P^o \tag{3.65}$$

with the following coefficients:

$$\begin{aligned} a_P &= a_W + a_E + a_P^o, \\ a_P^o &= \rho\frac{\Delta x}{\Delta t}, \\ a_W &= \frac{\Gamma_w}{\Delta x}, \\ a_E &= \frac{\Gamma_e}{\Delta x}. \end{aligned} \tag{3.66}$$

An obvious advantage with the fully implicit method is that the coefficients in Eq. (3.66) are always positive, giving unconditional boundedness. Thus, there is no strict upper limit in the choice of the appropriate time step. However, there is usually an optimal number of iterations to be used in each time step, which will restrict its value. Solving Eq. (3.65) with the coefficients in Eq. (3.66) with the time step $\Delta t = 0.5$ s gives a similar plot to that in the explicit case, see Figure 3.14. As in the explicit case,

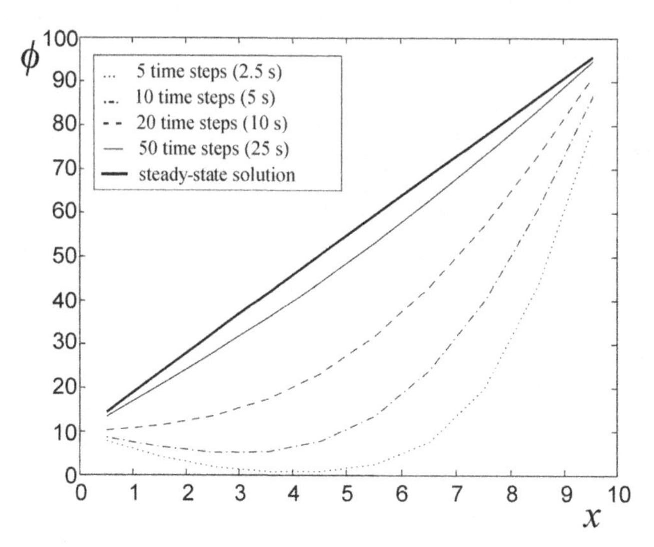

Figure 3.15 A plot of molar concentration as a function of the distance at different times. The time step is 0.5 s and the discretization method is Crank–Nicolson. The initial condition was $c = 0$ everywhere.

the results are reasonable. Diffusion takes care of the transport of species A into the domain.

The Crank–Nicolson method

In the Crank–Nicolson method, both the old value and the new value of ϕ are used to estimate the integral value of ϕ over a time step $\theta = 0.5$. Equation (3.60) then becomes

$$a_{\mathrm{P}}\phi_{\mathrm{P}} = a_{\mathrm{W}}\left[\frac{1}{2}\phi_{\mathrm{W}} + \frac{1}{2}\phi_{\mathrm{W}}^{\mathrm{o}}\right] + a_{\mathrm{E}}\left[\frac{1}{2}\phi_{\mathrm{E}} + \frac{1}{2}\phi_{\mathrm{E}}^{\mathrm{o}}\right] + \left[a_{\mathrm{P}}^{\mathrm{o}} - \frac{1}{2}a_{\mathrm{W}} - \frac{1}{2}a_{\mathrm{E}}\right]\phi_{\mathrm{P}}^{\mathrm{o}} \quad (3.67)$$

with the coefficients

$$\begin{aligned} a_{\mathrm{P}} &= \frac{1}{2}(a_{\mathrm{W}} + a_{\mathrm{E}}) + a_{\mathrm{P}}^{\mathrm{o}}, \\ a_{\mathrm{P}}^{\mathrm{o}} &= \rho\frac{\Delta x}{\Delta t}, \\ a_{\mathrm{W}} &= \frac{\Gamma_{\mathrm{w}}}{\Delta x}, \\ a_{\mathrm{E}} &= \frac{\Gamma_{\mathrm{e}}}{\Delta x}. \end{aligned} \quad (3.68)$$

Solving Eq. (3.67) together with its coefficients in Eq. (3.68) yields with $\Delta t = 0.5$ s the results shown in Figure 3.15.

Just as in the previous cases, the results look well shaped. Before moving on, we note that

$$\Delta t < \frac{\rho(\Delta x)^2}{\Gamma} \quad (3.69)$$

must be fulfilled in order to avoid unbounded solutions.

3.11.2 Conclusions on the different time discretization methods

As when the different discretization methods for spatial discretization were presented and evaluated in earlier sections of this chapter, accuracy, boundedness and transportiveness will be discussed here. The explicit method as well as the fully implicit method are first-order accurate, which we state here, without proof,

$$\phi = \phi^{\text{ex}} + O(\Delta t)\,,$$
$$\phi = \phi^{\text{im}} + O(\Delta t)\,.$$

This follows from the fact that the value of ϕ in each time step is taken from *either* the old value *or* the new one. The Crank–Nicolson method is second-order accurate, or in mathematical terms,

$$\phi = \phi^{\text{C-N}} + O\left[(\Delta t)^2\right].$$

The Crank–Nicolson method has major resemblances to the central-differencing scheme for spatial discretization, cf. Eqs. (3.13) and (3.57) with $\theta = 0.5$. A reduction of the time-step size thus gives a more significant improvement in the results if the Crank–Nicolson method is used than for first-order methods. There also exist many higher-order numerical schemes for time discretization. Figure 3.15 shows a simulation using the Crank–Nicolson method.

As concluded earlier, the explicit method is only conditionally bounded. This imposes a constraint on the choice of the time step, cf. Eq. (3.64). If this criterion is violated, the solution often diverges. On the other hand, being without the need to sub-iterate, the explicit method is *faster* than the other two.

The fully implicit method is unconditionally bounded. This is a great advantage with the method, and, in most commercial CFD codes, the fully implicit method is the default method for time discretization.

The Crank–Nicolson method is only conditionally bounded, cf. Eq. (3.69). This constraint is, however, not as restrictive as in the explicit method; it differs by a factor of 2.

The time step is determined by the Courant number (CFL). Since the time step is determined by the fluxes in each cell and the general rule is that the time step should be shorter than the time it takes to transport past the cell,

$$\Delta t < \text{CFL}\,\min\left(\frac{\rho(\Delta x)^2}{\Gamma}, \frac{\Delta x}{U}\right), \qquad (3.70)$$

where CFL $= 1$ for an explicit solver and CFL $= 5$ or higher for a fully implicit solver. A fully implicit solver should be stable for all time steps, but, due to nonlinearities in the equations, it is recommended that one start with a low Courant number (5) and increase it during the iterations.

Among the most recent transient solvers, the use of the *method of lines* is very common. This method reformulates the transient PDE into a transient ordinary differential equation

(ODE) by discretizing the spatial coordinates only. Then an ordinary ODE solver is used to solve the equations.

Questions

(1) The first term in Eq. (3.53) can be seen as a convective term for time. Why is there no diffusion term for time? What would such a term look like?
(2) What does Figure 3.12 tell us? What consequences does it enforce on the dependent variables?

3.12 Meshing

Finally, a few words will be said about meshing. So far, it has been assumed in all the examples that a proper mesh has been provided. However, this is generally not the case. Thus, *meshing* becomes an important part of the CFD engineer's work. Bad meshes often give numerical problems and bad results.

3.12.1 Mesh generation

There are several commercial software packets for mesh generation. To generate a mesh is in general a very complex procedure, and only some of the basics will be mentioned here.

Traditionally, grids have been divided into *structured grids* and *unstructured grids*. The structured grids are built up from quadrilateral elements, i.e. four-edged elements, but not necessarily rectangles, in 2D and hexahedra (elements with six faces) in 3D. Indexing and finding neighbouring cells is very easy with these elements. CFD programs using structured grids are usually faster and require less memory than programs using unstructured grids. Unstructured grids are built from different elements, quadrilateral and triangular elements in 2D and tetrahedra, hexahedra, pyramids, prisms and even dodecahedra in 3D. Usually, a structured grid will require less memory and have better numerical properties. However, it is not always possible to mesh complex geometries using structured mesh and today most solvers can handle both structured and unstructured grids. Figure 3.16 shows typical building elements for meshing in 2D and 3D.

Most meshing programs allow the user to draw the geometry in the program itself, but most commonly the geometry is imported from a general CAD program. The geometry is built from single points, lines or 2D and 3D shapes, e.g. rectangles, circles, boxes or cylinders. The points can be combined to make lines, and lines to 2D shapes etc., so any geometrical shape can be formed. It is then possible to unite, subtract or intersect these shapes and form new shapes.

Elements at the walls must be handled carefully. The angle between the wall and the grid line should be close to 90°. The easiest way to obtain this is to start the meshing by forming a regular mesh that is built from the surface. When a fine resolution is needed

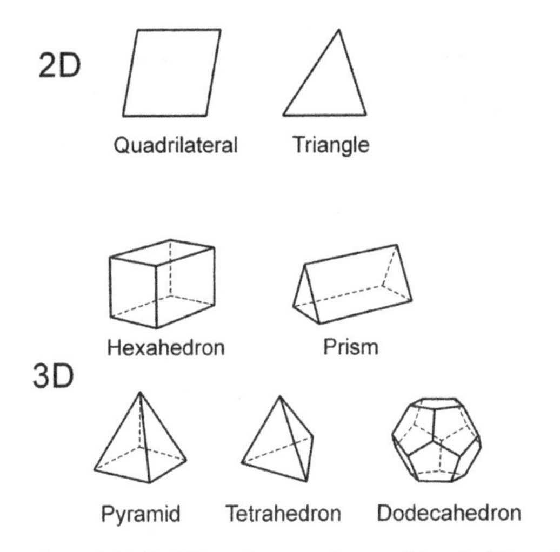

Figure 3.16 Building elements for meshing in 2D and 3D.

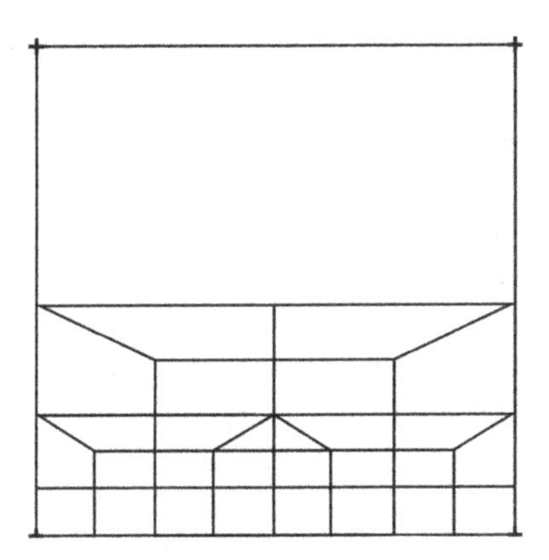

Figure 3.17 A boundary-layer mesh.

on the boundary, a mesh that increases in size with distance from the wall can be formed as shown in Figure 3.17.

There are many other properties of the mesh that should be taken into account in order to produce an accurate solution. For example, it can be shown that, if adjacent cells are very different in size (or volume), the numerical error will increase. It can also be shown that a high skewness of the cells can lead to instabilities and lower accuracy. The optimal situation is to have cells that have 90° corners and edges of equal length. For example, optimal quadrilateral meshes will have vertex angles close to 90°, while triangular meshes should preferably have angles close to 60° and have all angles less than 90°. In tetrahedral meshes the angles should be kept between 40° and 140°. One way to decrease the number of cells and enhance the convergence rate is to stretch the cells along a coordinate axis. This will increase their aspect ratio, but this is usually acceptable as long as the aspect ratio is kept below 5. For example, in long thin channels,

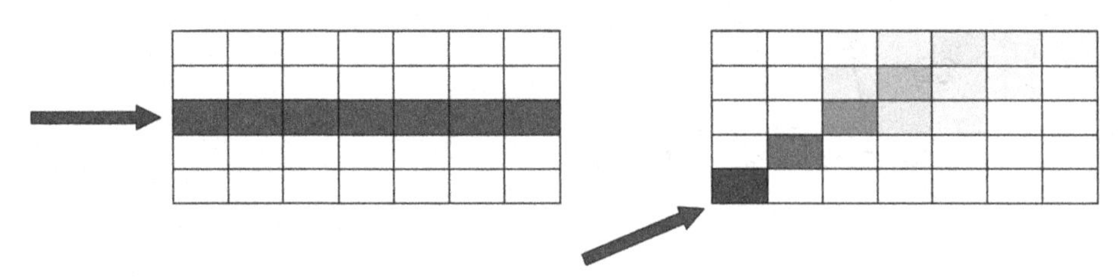

Figure 3.18 The effect of numerical diffusion. In the cases shown, the grid is aligned with the flow (left), and has inclined alignment (right).

it could be a good idea to use long cells in the streamwise direction since there are hardly any gradients in that direction.

3.12.2 Adaptation

It has already been concluded that the distance between adjacent cells, namely the grid spacing, Δx, plays an important role in determining the accuracy of the solution. The denser the mesh, the more accurate the solution. If consistency is upheld, the solution to the discretized problem tends to the exact solution of the set of PDEs as the grid spacing is reduced infinitely. It has also been shown that the required grid spacing is related to the order of the discretization scheme.

In many cases, a good idea could be to use different grid spacings in different regions of the grid. In this way, it is possible to resolve some areas to a very large extent, while other areas are not resolved any more than is necessary in order to avoid divergence. Boundary-layer flows, for example, generally require a very dense mesh close to the wall, while the flow far from the wall does not have to be resolved in detail. This is a consequence of the fact that areas with large gradients normally contain larger errors and thus have to be better resolved. In order to know where the largest gradients are, a simulation has to be performed or the user has to trust his or her intuition and previous experience. Then grid refinement/coarsening is done in the appropriate areas. A new simulation is then performed. If necessary, further refinement/coarsening can be done. Most CFD software has a built-in grid adaptation/coarsening algorithm.

3.12.3 Numerical diffusion

Care must be taken when choosing the directions of the cell axes. In a structured grid it is common to dispose the cells in a manner parallel to the flow. To illustrate what can happen if the cells are disposed in an inclined manner, consider the following (see also Figure 3.18).

A liquid stream with a non-diffusing species is flowing with a constant velocity. If the cells are parallel to the flow, there will be no transfer of the species in the wall-normal direction. This would of course be the case in reality as well. However, if the cells are not parallel to the flow, there will be a transport of species due to the fact that the value of the species entity is the same throughout the cell. Hence, there will be transport due to the discretization. This is called *numerical*, or *false*, *diffusion*. By refining the mesh it is possible to reduce the effect of numerical diffusion. The use of higher-order discretization schemes will also reduce this effect.

3.13 Summary

The basics of the numerics behind a CFD simulation have now been presented. It is important to keep in mind that this has only been a short introduction. At this stage the reader should be able to understand the fundamentals of commercial CFD software. The reader should also be able to understand some of the problems that can occur and how they can be avoided.

Questions

(1) Explain what is meant by adaptation.
(2) Describe the principle of multigrid methods.
(3) What is meant by the Courant number?
(4) Describe the difference between implicit and explicit methods.
(5) What is numerical diffusion and how can it be minimized?

4 Turbulent-flow modelling

This chapter provides an insight into the physical nature of turbulence and the mathematical framework that is used in numerical simulations of turbulent flows. The aim is to explain why turbulence must be modelled and how turbulence can be modelled, and also to explain what is modelled with different turbulence models. In addition, the limitations of the turbulence models are discussed. The intention is to give you such an understanding of turbulence modelling that you can actually suggest appropriate turbulence models for different kinds of turbulent flows depending upon their complexity and the required level of description.

4.1 The physics of fluid turbulence

Turbulence is encountered in most flows in nature and in industrial applications. Natural turbulent flows can be found in oceans, in rivers and in the atmosphere, whereas industrial turbulent flows can be found in heat exchangers, chemical reactors etc. Most flows encountered in industrial applications are turbulent, since turbulence significantly enhances heat- and mass-transfer rates. In industry a variety of turbulent multiphase flows can be encountered. Turbulence plays an important role in these types of flows since it affects processes such as break-up and coalescence of bubbles and drops, thereby controlling the interfacial area between the phases. Thus, turbulence modelling becomes one of the key elements in CFD.

In order to determine which turbulence-modelling approach is best suited to a particular application, we need to understand the limitations of various turbulence models. This insight requires a certain level of understanding of fluid turbulence. Probably everyone has an intuitive understanding of what turbulence is, since turbulence is encountered daily. From everyday experiences e.g. mixing coffee and milk, we know that turbulence increases the disorder in a fluid, resulting in an efficient mixing of fluid elements. As has already been mentioned, turbulence significantly affects mass-, momentum- and heat-transfer rates. High mass- and heat-transfer rates are positive and crucial aspects of many processes. In contrast, increased momentum transfer is usually undesirable, since it results in higher skin friction and hence drag. In other words, turbulence is an unavoidable feature of many chemical processes because high throughputs imply turbulent conditions. In this section we will take a closer look at what characterizes turbulence.

According to Hinze [5], turbulence can be characterized as follows:

Turbulent fluid motion is an irregular condition of flow in which the various quantities show random variation with time and space coordinates, so that statistically distinct average values can be discerned.

Indeed, turbulence is a state of fluid flow that can be characterized by random and chaotic motions. However, we need a deeper understanding of fluid turbulence before we can discuss turbulence modelling and understand the limitations of particular turbulence models.

4.1.1 Characteristic features of turbulent flows

Since it is difficult to give an exact definition of turbulence, we will look into several characteristic features of fluid turbulence. Observations of whirling smoke above a cigarette or other turbulent motions clearly show that large vortices in such motions are unstable and break up into smaller vortices. From such an observation it can also be seen that the flow becomes less chaotic far away from the cigarette. These two observations tell us that turbulence is a decaying process whereby large turbulent structures break up into subsequently smaller and smaller structures until the flow becomes laminar. Actually, turbulence dies out rather quickly if energy is not continuously supplied. There are more characteristic features of turbulence that can be identified by studying turbulent flows. Tennekes and Lumley [6] did not give a precise definition of turbulence; instead they listed the most important features of turbulence. These characteristic features are as follows.

(1) Irregularity. Turbulent flows are irregular, random and chaotic, and consist of a wide range of length scales, velocity scales and timescales. The large-scale motions in turbulent flows are usually referred to as large eddies or large vortices. A turbulent eddy is a turbulent motion that over a certain region is at least moderately coherent. The region occupied by a large turbulent eddy can also contain smaller turbulent eddies. This means that different scales coexist and smaller scales exist inside large scales. In turbulent flows the largest scales are bounded by the geometry of the flow, whereas the smallest scales are bounded by viscosity. The smallest eddies are generally several order of magnitudes smaller than large eddies. While eddies move they stretch, rotate and break up into two or more eddies. The irregularity of turbulent flows and the wide range of length scales and timescales make a deterministic approach to turbulence simulation very difficult. Statistical models are therefore frequently used in practical engineering simulations. An instantaneous representation of turbulent eddies in a pipe flow is shown in Figure 4.1.

Passages of large and small eddies through a certain point in a turbulent-flow field induce irregular velocity fluctuations. A point measurement of the instantaneous velocity in a turbulent flow yields a fluctuating velocity field similar to the one shown in Figure 4.2. Note that passages of large eddies through this point induce

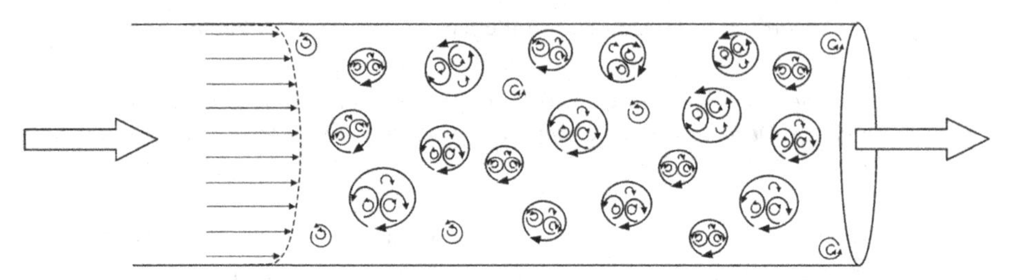

Figure 4.1 Large- and small-scale turbulent structures in a turbulent pipe flow.

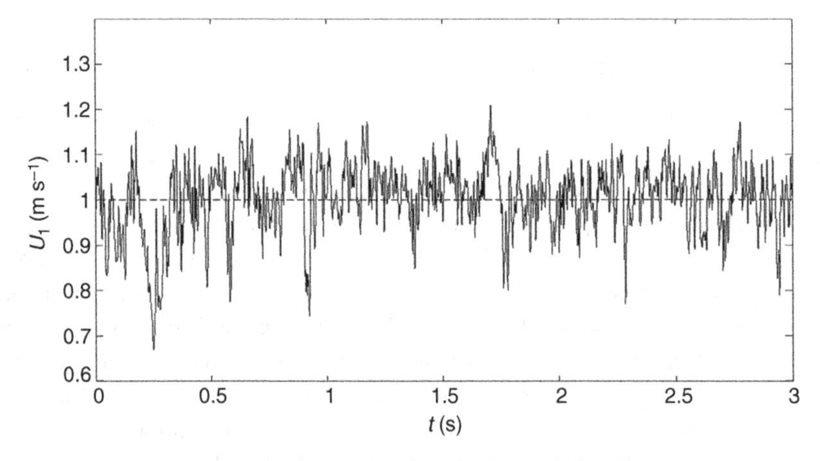

Figure 4.2 Instantaneous velocity at a certain point in a turbulent flow.

fluctuations of large amplitude and low frequency, whereas small eddies induce fluctuations of small amplitude and high frequency.

(2) Diffusivity. Probably the most important characteristic of turbulence is that it is diffusive. The turbulent diffusive transport is due to the chaotic motions in the flow and allows faster mixing rates of species, momentum and energy than would be allowed by the molecular diffusion alone. These rates are generally several orders of magnitude larger than the rate of molecular diffusion. Since turbulence is a 3D phenomenon, the turbulent transport occurs in all three dimensions. A simplified illustration of this turbulent transport is shown in Figure 4.3. In this figure two fluid elements are transported perpendicular to the streamlines. In contrast to laminar flow, this transport occurs even though the mean velocity component is zero in the y direction.

(3) Instability at large Reynolds numbers. Turbulence arises due to instabilities occurring at high Reynolds numbers. From a physical point of view this happens when the timescale for viscous damping of a velocity fluctuation is much larger than the timescale for convective transport. An extended discussion of this process is given in Section 4.1.3. From a mathematical point of view this can also be seen in the Navier–Stokes equation, where the nonlinear convective term becomes more important

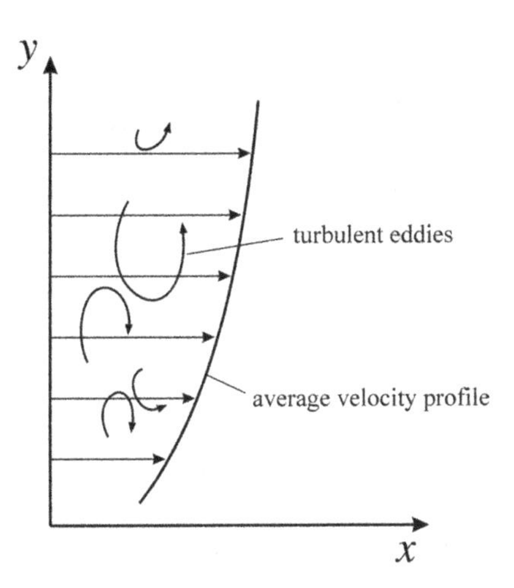

Figure 4.3 Transport due to turbulent convection.

than the viscous term with increasing Reynolds number. In dimensionless form the Navier–Stokes equation reads

$$\frac{\partial \widetilde{U_i}}{\partial \tilde{t}} + \widetilde{U_j} \frac{\partial \widetilde{U_i}}{\partial \tilde{x}_j} = -\frac{\partial \widetilde{P}}{\partial \tilde{x}_i} + \frac{1}{Re} \frac{\partial^2 \widetilde{U_i}}{\partial \tilde{x}_j \partial \tilde{x}_j}. \tag{4.1}$$

Thus the tendency towards instability, which is damped by viscosity, increases with the Reynolds number. Turbulent flows appear random in time and space, and are not experimentally reproducible in detail. Note that the nature of turbulence is random even though the Navier–Stokes equations are deterministic. In any turbulent flow there are unavoidable perturbations in initial conditions, boundary conditions and material properties. Turbulent fields display an acute sensitivity to such perturbations. This means that turbulence is stochastic even though the Navier–Stokes equations are deterministic.

(4) Three-dimensional structures. Turbulence is intrinsically 3D. The reason for this is that mechanisms such as vortex stretching and vortex tilting cannot occur in two dimensions (see the discussion in Section 4.1.6). Nevertheless, turbulent flows can be 2D in a statistical sense, hence 2D simulations of turbulent flows can be performed. Actually most turbulence modelling applied to practical engineering applications is based on models in which the 3D fluctuations are filtered out, thus not resolving the turbulent fluctuations, but do resolve the coupling between the fluctuations and the mean flow field.

(5) Dissipation of turbulent kinetic energy. In all turbulent flows there is a flux of energy from the largest turbulent scales to the small scales. At the smallest scales the turbulent kinetic energy is dissipated into heat due to viscous stresses. This flux of energy is commonly referred to as the energy cascade. The idea of the energy cascade is that kinetic energy enters the turbulence at the largest scales at which energy is extracted from the mean flow. By inviscid processes this energy is then transferred to smaller and smaller scales. The reason for this energy flux is that large eddies are

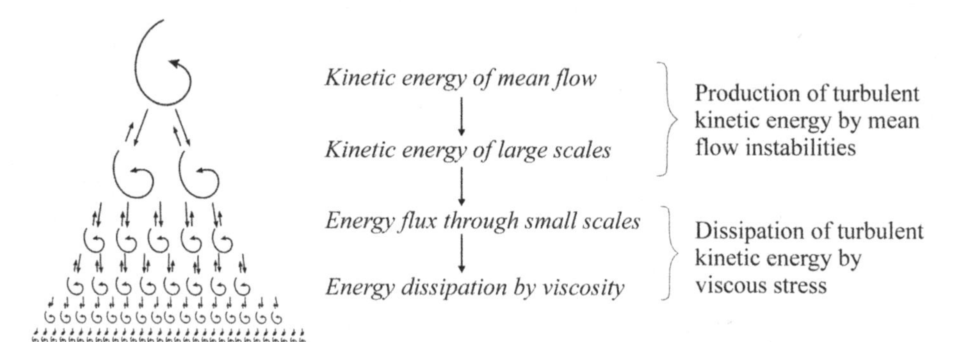

Figure 4.4 Energy flux from large to small scales.

unstable and break up into smaller eddies, thereby transferring the energy to smaller scales. These smaller eddies undergo similar break-up processes and transfer the energy to yet smaller eddies. On the smallest scale we find the dissipative eddies, whose energy is dissipated into heat by viscous action due to molecular viscosity. This energy flux is illustrated in Figure 4.4. A consequence of the dissipation is that turbulence decays rapidly if no energy is supplied to the system. In chemical processes energy is supplied to systems in numerous ways. In a turbulent pipe flow, the energy is supplied by the pump, whereas in a stirred-tank reactor the energy is supplied by the impeller. The amount of energy supplied to these systems can be determined from the pressure drop and torque, respectively. It should be pointed out that the total energy input equals the sum of energy losses due to energy dissipation in the fluid and at the walls.

(6) Continuum. Turbulence is a continuum phenomenon in which even the smallest scales of turbulence are much larger than the molecular length scale. The motion of fluids is therefore described by the conservation equations for mass and momentum conservation supplemented by initial and boundary conditions.

(7) Turbulent flows are flows. Turbulence is a feature of a flow, not a fluid. This means that all fluids can be turbulent at high enough Reynolds number.

4.1.2 Statistical methods

More than a century ago, Reynolds introduced statistical averaging methods for turbulent flows. Statistical methods still remain crucial in the theory of turbulence and in turbulence modelling. These methods are used to study mean values of flow properties in space and time. By using statistical methods, the need for information about the flow is reduced and the flow description is considerably simplified. In this book we will focus on single-point statistics. Even though one-point quantities do not encompass the full statistics of the flow, they still include many of the more important measures of turbulent flows such as the mean flow velocity, $\langle U_i \rangle$, and the turbulent kinetic energy per unit mass, $\langle u_i u_i \rangle / 2$. Spectral analysis is an example of another method that better describes the flow. With this method, it is even possible to capture the contributions of different scales to the

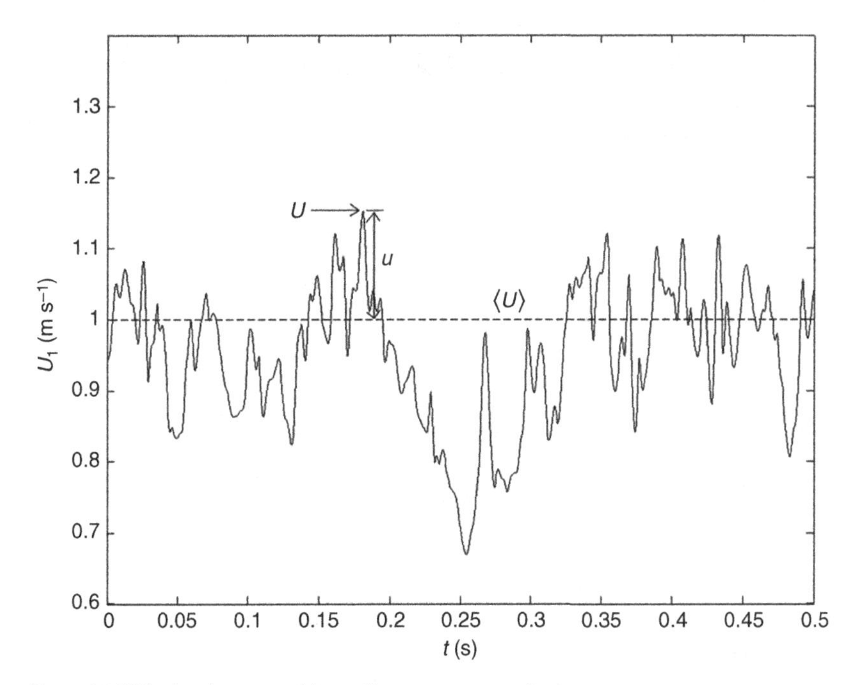

Figure 4.5 Velocity decomposition of instantaneous velocity.

overall energetics of turbulence. For more information about spectral analysis and other advanced methods the reader is referred to textbooks on turbulence [7–12].

One-point measurement of the instantaneous velocity reveals that passage of small eddies induces fluctuations of small amplitude and high frequency, whereas passage of large eddies induces fluctuations of large amplitude and low frequency. Frequencies ranging from 1 Hz to 10 000 Hz are commonly observed in air and water flows. An example of such a measurement is shown in Figure 4.5. In this figure it can be seen that fluctuations with different frequencies exist. In the one-point statistics, the instantaneous velocity at any particular position and time, $U_i(x, t)$, is divided into an average and a fluctuating part. The average represents the mean velocity, whereas the fluctuating parts is usually interpreted as representing the turbulence. Decomposition of the instantaneous velocity into its mean and fluctuating parts is then represented as

$$U_i = \langle U_i \rangle + u_i, \tag{4.2}$$

where the average velocity is defined as

$$\langle U_i \rangle = \frac{1}{2T} \int_{-T}^{T} U_i \, dt. \tag{4.3}$$

The timescale used in this filtering operation is chosen so that the instantaneous variables are averaged over a time period that is large compared with the turbulent timescales but small compared with the timescale of the mean components. Thus, turbulent flows can be considered to consist of randomly varying components superimposed on a mean motion. This decomposition is known as Reynolds decomposition since Reynolds first introduced the concept in the analysis of turbulent flows. The intensities of velocity fluctuation in different directions can be measured in terms of the turbulent kinetic energy per unit mass

$$k = \frac{1}{2}\langle u_i u_i \rangle = \frac{1}{2}\sum_{i=1}^{3}\langle u_i^2 \rangle = \frac{1}{2}\left(\langle u_1^2 \rangle + \langle u_2^2 \rangle + \langle u_3^2 \rangle\right). \quad (4.4)$$

Let us now define what we mean by turbulent kinetic energy. The kinetic energy of the fluid (per unit mass) at a specific point in time is given by

$$E = \frac{1}{2}U_i U_i. \quad (4.5)$$

The mean of the kinetic energy can be decomposed into two parts,

$$\begin{aligned}\langle E \rangle &= \frac{1}{2}\langle(\langle U_i \rangle + u_i)(\langle U_i \rangle + u_i)\rangle = \frac{1}{2}(\langle U_i \rangle\langle U_i \rangle + \langle u_i u_i \rangle) \\ &= \frac{1}{2}\langle U_i \rangle\langle U_i \rangle + k = \overline{E} + k, \end{aligned} \quad (4.6)$$

where $\overline{E}$ is the kinetic energy of the mean flow and k is the turbulent kinetic energy per unit mass, $\frac{1}{2}\langle u_i u_i \rangle$.

To solve the Navier–Stokes equations, the pressure is decomposed in a similar way:

$$P = \langle P \rangle + p. \quad (4.7)$$

For incompressible fluids there is no need for decomposition of any other quantities than velocity and pressure. However, for compressible fluids the density must also be decomposed.

Definition of various statistical symmetries

A steady or stationary turbulent flow is defined as one whose statistical properties do not change with time. Homogeneity implies that, given any number of different spatial points and times, the statistics will remain unchanged if all positions are shifted by the same constant displacement. In other words, the statistics of the flow are invariant under translation. If the statistics are also invariant under rotation and reflection, the flow is isotropic. Isotropy is therefore a stricter criterion than homogeneity. Assumptions of homogeneous and isotropic turbulence are often made in theoretical studies, since these assumptions simplify equations and analysis. We can summarize the statistical symmetries as

- statistically stationary, if all statistics are invariant under a shift in time;
- statistically homogeneous, if all statistics are invariant under a shift in position (translation);
- statistically isotropic, if all statistics are invariant under rotations and reflections.

Even though homogeneous turbulence is rarely encountered in practice it is still possible to produce laboratory flows close to homogeneity. Grid-generated turbulence is an example of approximately homogeneous turbulence. These flows are homogeneous in

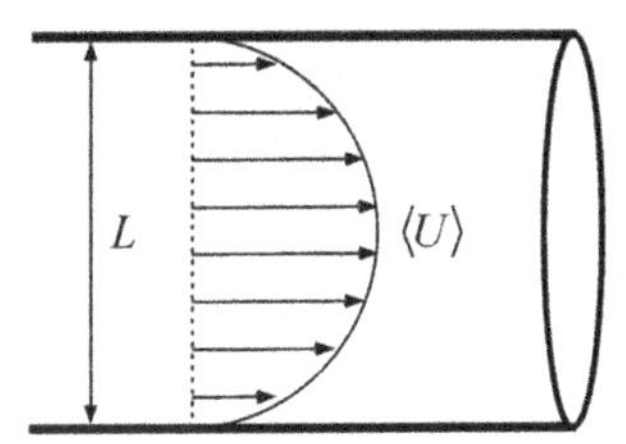

Figure 4.6 Pipe flow.

the sense that the variations of the statistical properties take place over distances large compared with the length scales of the turbulence itself. Grid turbulence in wind tunnels is therefore extensively studied in academia. Knowledge originating from experiments of this type has been widely used to develop and validate turbulence models.

4.1.3 Flow stability

Recall that the Reynolds number is often used to determine the transition between laminar and turbulent flows. The Reynolds number is the ratio of inertial force to viscous force,

$$Re = \frac{\langle U \rangle L}{\nu}. \tag{4.8}$$

Hence, increasing the Reynolds number represents a relative increase in the inertial force in relation to the viscous force. Transition from laminar to turbulent flow is related to the presence of a disturbance and its amplification. Thus the damping or amplification of a small velocity fluctuation in laminar flow determines the stability of the flow. If the disturbance is damped, the flow remains stable and laminar. On the other hand, if it is amplified, the flow becomes unstable and this can lead to turbulence through transition. Transition from laminar to turbulent flow occurs when the time taken to equilibrate with a wall due to diffusive transport is much larger than the time for convective transport. Consider a pipe flow, illustrated in Figure 4.6, characterized by a pipe diameter L, an average velocity $\langle U \rangle$ and a velocity fluctuation u.

Turbulent eddies are created in the near-wall region and, when there is a lack of damping by viscosity, they can move across the pipe, keeping the flow turbulent. If u is the eddy velocity and L is proportional to the nearest surface that can damp the turbulent fluctuations, the necessary time for this transport is

$$t_c = L/u. \tag{4.9}$$

During this time period viscosity acts on a distance

$$l = (t_c \nu)^{1/2}. \tag{4.10}$$

This means that viscosity is able to damp velocity fluctuations of scale smaller than l. The condition for turbulent flow is then given by

$$l < L, \tag{4.11}$$

or

$$l = (t_c \nu)^{1/2} < L. \tag{4.12}$$

Substitution of Eq. (4.9) into Eq. (4.12) leads to the condition

$$\frac{uL}{\nu} \gg 1. \tag{4.13}$$

Note that the velocity fluctuations, u, are usually much smaller than the average velocity in the pipe, $\langle U \rangle$. Thus, the criterion for turbulent flow in pipes in terms of the average velocity is given by

$$Re = \frac{\langle U \rangle L}{\nu} \gg 1. \tag{4.14}$$

This means that, for high-Reynolds-number flows, $Re \gg 1$, viscosity cannot damp velocity fluctuations and the flow becomes turbulent. The first systematic studies regarding the transition to turbulence were carried out by Osborne Reynolds more than a century ago. Reynolds studied flow in glass tubes and obtained a critical Reynolds number 2100 at which the flow ceased to be laminar. By eliminating disturbances at the inlet of the tube, it has been possible to maintain laminar flow even for Reynolds numbers one order of magnitude larger than that.

Experimental studies have shown that the following Reynolds-number criteria can be used to predict the transition from laminar to turbulent flow.

(1) For internal flows,
   - pipe flow, $Re > 2100$;
   - flow between parallel plates, $Re > 800$.
(2) For external flows,
   - flow around a sphere, $Re > 350$.
(3) Boundary layers,
   - flow along surfaces, $Re > 500\,000$.

Note that different characteristic lengths are used in the definitions of the Reynolds numbers above. It should also be noted that surface roughness affects the transition as well.

4.1.4 The Kolmogorov hypotheses

In 1941, Kolmogorov stated three prominent hypotheses that are of fundamental importance for the understanding of turbulence. The first hypothesis concerns the isotropy of small-scale turbulent motions. Kolmogorov argued that there are reasons to expect that at high Reynolds number the directional information is lost in the chaotic scale-reduction process. Whereas the anisotropic large-scale structures, l_0, depend on geometry and boundary conditions, it is assumed that the small scales have small timescales and that these motions are statistically independent of the large-scale turbulence and of the mean flow. In other words, somewhere in the process whereby turbulent eddies are reduced in size all directional information is lost. On this scale turbulence is statistically isotropic and these scales are therefore independent of the geometry.

Kolmogorov's hypothesis of local isotropy:

At sufficiently high Reynolds numbers, the small scales of turbulent motions, $l \ll l_0$, *are statistically isotropic.*

Kolmogorov also argued that the statistics of small-scale motions are universal, i.e. similar in every high-Reynolds-number flow. This argument forms the basis for the second hypothesis. For the small-scale motions, the isotropic scales ($l < l_{EI}$), the transfer of energy to successively smaller scales and energy dissipation are the dominant processes. This leads to the conclusion that the energy-transfer rate and the kinematic viscosity are the two important parameters determining the statistics of the small-scale motions. In other words, turbulent structures much smaller than the anisotropic structures are universal, being solely determined by the energy-dissipation rate, ε, and the viscosity, ν. This hypothesis is known as 'Kolmogorov's first similarity hypothesis'.

Kolmogorov's first similarity hypothesis (dissipative range):

In every turbulent flow at sufficiently high Reynolds number, the statistics of the small-scale motions, $l < l_{EI}$, *have a universal form that is uniquely determined by viscosity,* ν, *and dissipation rate,* ε.

The characteristic length scale, velocity scale and timescale in the dissipative range are thus given by the energy-dissipation rate, ε, and the viscosity, ν. These scales are also known as the Kolmogorov scales. The Kolmogorov scale, η, characterizing the size of the smallest turbulent eddies, which is the scale for dissipation of turbulent kinetic energy, is

$$\eta = \left(\frac{\nu^3}{\varepsilon}\right)^{1/4} \tag{4.15}$$

and the Kolmogorov velocity scale is given by

$$u_\eta = (\varepsilon\nu)^{1/4}. \tag{4.16}$$

The Kolmogorov timescale, τ_η, for viscous dissipation is given by

$$\tau_\eta = \left(\frac{\nu}{\varepsilon}\right)^{1/2}. \tag{4.17}$$

By definition Kolmogorov's length and velocity scales give a Reynolds number equal to 1. Thus, the smallest-scale motion of turbulence is laminar, being solely determined by viscous forces.

Within the range $l < l_{EI}$ the timescales are small compared with the timescales for the large eddies. This means that the small eddies can adapt quickly to maintain a dynamic equilibrium with the energy-transfer rate imposed by the large eddies. Since the statistics of the small-scale motions are universal, this range $l < l_{EI}$ is often referred to as the universal equilibrium range.

Kolmogorov also stated a 'second similarity hypothesis' in which it is assumed that for a special range of structures within the universal equilibrium range viscosity plays

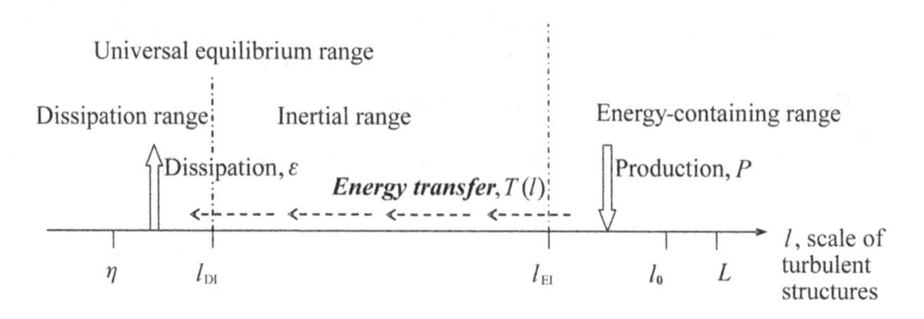

Figure 4.7 The cascade of turbulence energy on a logarithmic scale.

a negligible role in the motions. This means that only the energy-dissipation rate, ε, determines the statistics of the motions in this range. This is the so-called inertial range.

Kolmogorov's second similarity hypothesis (inertial range):

In every turbulent flow at sufficiently high Reynolds number, the statistics of the motions of scale l in the range $\eta \ll l \ll l_0$ have a universal form that is uniquely determined by ε and independent of ν.

On introducing a length scale $l_{\mathrm{DI}} = 60\eta$, the inertial range is given by $l_{\mathrm{DI}} < l < l_{\mathrm{EI}}$. Note that the length scale l_{DI} splits the universal equilibrium range into two subranges. These subranges are the inertial range, $l_{\mathrm{DI}} < l < l_{\mathrm{EI}}$, and the dissipative range, $l < l_{\mathrm{DI}}$.

Thus, according to the two similarity hypotheses, the motions in the inertial subrange are determined solely by inertial effects, whereas the motions in the dissipation range experience significant viscous effects. Figure 4.7 shows the various length scales and ranges.

4.1.5 The energy cascade

The basic idea in the energy-cascade theory, illustrated in Figure 4.7, is that there is a net flux of energy from large to small eddies. The idea of energy transfer from large to subsequently smaller scales was introduced by Richardson in 1922. In the energy cascade there is a source of turbulent energy, P, at the largest scales. On the largest scales energy is extracted from the mean flow instabilities. This means that, if there were no flow, no fluctuations would be sustained; hence the mean flow drives the fluctuations. Thus, the energy cascades start with energy transfer from the mean flow to the largest eddies. It is assumed that large eddies contain the largest part of the energy and contribute little to the energy dissipation. It is also assumed that the largest eddies in turbulent flows do most of the transport of momentum and other quantities. Since the continuous supply of energy can neither accumulate in large eddies nor just disappear, there must be a sink of energy. The sink of energy is twofold. Energy is finally transferred to heat on the bounding surfaces and it goes into heating up the fluid itself. Actually, the energy-cascade theory explains the latter mechanism. It is assumed that the flux of energy continues to successively smaller scales until the viscous stress becomes effective. On this scale, turbulent kinetic energy is dissipated as heat by molecular viscosity. The

Table 4.1 Correlations for various turbulence scales

Scale	Length	Time	Velocity
Large scale	$l = k^{3/2}/\varepsilon$	$\tau_l = k/\varepsilon$	$u_l = \left(\frac{2}{3}k\right)^{1/2}$
Smallest scale	$\eta = (\nu^3/\varepsilon)^{1/4}$	$\tau_\eta = (\nu/\varepsilon)^{1/2}$	$u_\eta = (\varepsilon\nu)^{1/4}$

energy-dissipation process is a result of viscous friction between layers of fluid moving at different velocities. An important consequence of the viscous stress is that it prevents the generation of infinitely small eddies. Hence, in all fluid flows there is a minimum scale of turbulent structures.

Discussions of the energy cascade often refer to the universal equilibrium range. This terminology is based on the argument that the small eddies will evolve much more rapidly than the large eddies. Thus eddies in the universal range can adjust so quickly to changes in external conditions that they can be assumed to be always in a state of local equilibrium. The transfer of energy in the cascade is given by

$$T_{\text{EI}} = T(l_{\text{EI}}) = T(l) = T_{\text{DI}} = T(l_{\text{DI}}) = \varepsilon. \tag{4.18}$$

In conclusion, the energy-cascade theory places production at the beginning and dissipation at the end of a sequence of processes. L and l_0 in Figure 4.7 are the flow scale and the scale of the largest eddies, respectively. The length scale l_0 is a measure of the largest turbulent eddies, which contain most of the turbulent kinetic energy. l_0 is given by

$$l_0 = \frac{k^{3/2}}{\varepsilon}. \tag{4.19}$$

The characteristic timescale for the large eddies, τ_l, is the time necessary to decrease a turbulent structure of size l_0, that is the 'eddy lifetime'. This timescale can also be seen as the timescale for transfer of turbulent kinetic energy from the scale l_0 to η, which means that it is a measure for the turbulence decay rate. τ_l is given by

$$\tau_l = \frac{k}{\varepsilon}. \tag{4.20}$$

Since k is defined as $\frac{1}{2}\left(u_1^2 + u_2^2 + u_3^2\right)$, the characteristic velocity scale for the large eddies is given by

$$u_l = \left(\frac{2}{3}k\right)^{1/2}. \tag{4.21}$$

As a rule of thumb, l_{EI} can be used as a demarcation between anisotropic and isotropic turbulent eddies, see Figure 4.7. The length scale of the anisotropic large eddies is then given by $l > l_{\text{EI}}$ and that for isotropic eddies by $l < l_{\text{EI}}$. This demarcation requires an approximation of l_{EI}, which is given by $l_{\text{EI}} \approx \frac{1}{6}\, l_0$. The relations for the various scales of turbulent motions are summarized in Table 4.1.

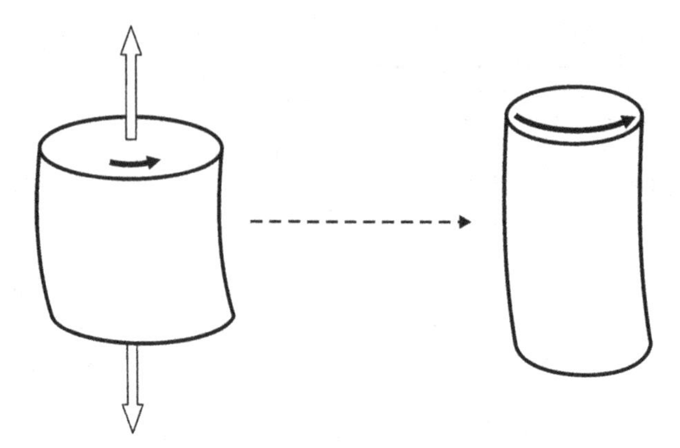

Figure 4.8 Stretching of a vortex tube concentrates vorticity on progressively smaller scales.

From the correlations in Table 4.1 we can determine the smallest turbulent scales (Kolmogorov scales) in a 100-W kitchen mixer. Assume that the mixer is filled with 1 litre of water ($\nu = 10^{-6}\ \mathrm{m^2\,s^{-1}}$ and $\rho = 10^3\ \mathrm{kg\,m^{-3}}$) and that all energy put into the system is dissipated homogeneously in the fluid. Since $\mathrm{W} = \mathrm{J\,s^{-1}} = \mathrm{m^2\,kg\,s^{-3}}$, the energy-dissipation rate per unit mass, ε, is equal to $100\ \mathrm{m^2\,s^{-3}}$. Hence, the smallest turbulent length scale in the mixer is $\eta = (\nu^3/\varepsilon)^{1/4} = 10\ \mu\mathrm{m}$ and the characteristic timescale for these eddies is $\tau_\eta = (\nu/\varepsilon)^{1/2} = 0.1\ \mathrm{ms}$.

4.1.6 Sources of turbulence

Turbulent flows require a continuous supply of energy since turbulence is inherently dissipative. In the energy-cascade theory the source of energy enters at the largest scale, at which energy is extracted from the mean flow by the large-scale eddies. Whereas large eddies extract energy from the mean flow, small eddies are supplied with energy from the flux of energy from the large eddies. This energy transfer between eddies is assumed to be related to vortex stretching and the conservation of angular momentum when eddies are stretched. The interaction between vorticity and velocity gradients is an important mechanism to create and maintain turbulence. Two idealized mechanisms that result from this interaction are vortex stretching and vortex tilting. On average these mechanisms create smaller and smaller scales. Hence, stretching and tilting of vortices create and maintain turbulence at smaller scales. The stretching mechanism is illustrated in Figure 4.8. Stretching of a vortex tube means that the cross-section of the vortex tube decreases in size (for an incompressible fluid). In other words, the size of the vortex, as estimated from its transverse dimensions, becomes smaller. This argument shows how larger vortices or eddies in a turbulent fluid can give rise to smaller ones. The stretching work done by the mean flow on large eddies provides the energy which maintains turbulence. Smaller eddies are themselves stretched by somewhat larger eddies. In this way, energy transfers to progressively smaller scales. In this process the orientation of large eddies imposed by the mean flow is lost. Thus small scales will be

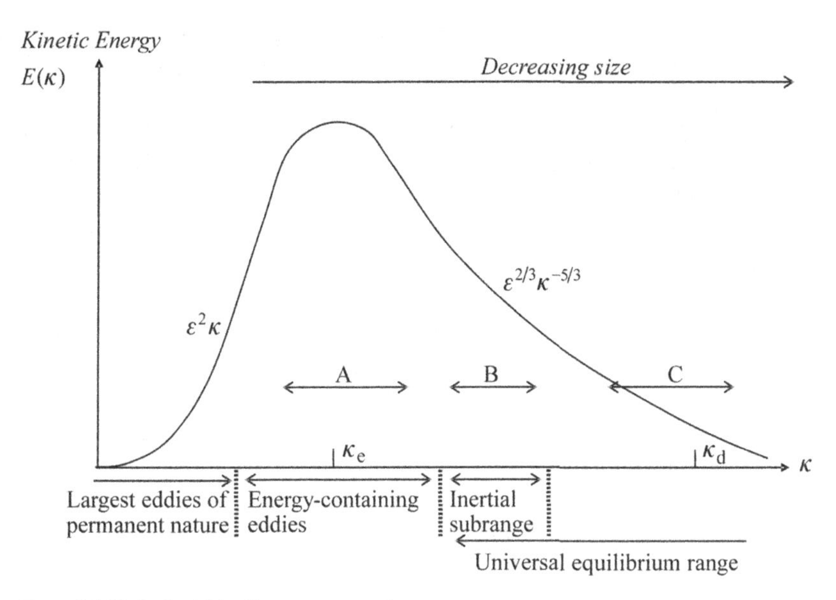

Figure 4.9 Turbulent kinetic energy spectrum.

isotropic. Turbulent flows dissipate energy as the viscous stresses act on the smallest scales. On this scale energy is dissipated into heat due to molecular viscosity.

4.1.7 The turbulent energy spectrum

At high Reynolds numbers, turbulent flows are characterized by the existence of a wide range of length scales that are bounded from above by the dimensions of the flow field and bounded from below by the diffusive action of molecular viscosity. One important tool for analysing the different regions of turbulence is the energy spectrum. It is common practice to use wave numbers instead of length scales. The dimension of a wave number is one over length, thus we can think of the wave number as inversely proportional to the eddy radius, i.e. $\kappa \propto 1/r$. This means that large wave numbers correspond to small eddies and small wave numbers to large eddies. Eddies with wave numbers in the region of κ_e, Figure 4.9, contain the largest part of the energy and contribute little to the energy dissipation. However, the small eddies, which are of very high frequency, in the region of κ_d, dissipate energy. Here, turbulent kinetic energy is dissipated into heat by molecular viscosity. The viscous stresses prevent the generation of eddies with higher frequency. In wave-number space, the energy of eddies from κ to $\kappa + d\kappa$ can be expressed as

$$E(\kappa)d\kappa. \tag{4.22}$$

The total turbulent kinetic energy, k, which is the sum of the kinetic energies of the three fluctuating velocity components, i.e. $k = \frac{1}{2}\langle u_i u_j \rangle$, is obtained by integrating over the whole wave-number space

$$k = \int_0^\infty E(\kappa)d\kappa. \tag{4.23}$$

The energy spectrum of fully developed homogeneous turbulence is thought to be composed of three distinct wave-number regions (see Figure 4.9).

A. In this region the large energy-containing eddies are found. These eddies interact with the mean flow and extract energy from the mean flow. The energy is transferred to slightly smaller scales and eventually into region B.
B. This region is the inertial subrange. In this region turbulent kinetic energy is neither produced nor dissipated. However, there is a net flux of energy through this region from A to C. The existence of this region requires that the Reynolds number is high.
C. This is the dissipative region where turbulent kinetic energy is dissipated into heat. Eddies in this region are isotropic and the scales are given by the Kolmogorov scales.

A spectral analysis of the turbulence scales often reveals a region where the distribution obeys the following relationship:

$$E(\kappa) = C_{\kappa}\varepsilon^{2/3}\kappa^{-5/3}, \qquad \frac{1}{l_0} \ll \kappa \ll \frac{1}{\eta}. \tag{4.24}$$

This equation is called the Kolmogorov spectrum law, or simply the $-5/3$ law. The equation states that, if the flow is fully turbulent, the energy spectra should exhibit a $-5/3$ decay. The region where this law applies is known as the inertial subrange and is the region where the energy cascade proceeds in local equilibrium. This law is often used in experiments and simulations (DNS, LES) to verify that the flow is fully turbulent. Note that the largest eddies, κ_e, contain most of the turbulence energy and are therefore responsible for most of the turbulent transport. Nevertheless, the small eddies are responsible for mixing on the small scales.

Questions

(1) Describe the process of transition from laminar to turbulent flow.
(2) Discuss how turbulence can be characterized.
(3) Explain the source of the energy supplied to eddies, why eddies have a lower limit in size and why the turbulent velocity field is isotropic within that range.
(4) Explain what is meant by the energy-containing range, the inertial subrange and the dissipative range in the energy spectrum.
(5) Explain what is meant by vortex stretching and relate it to the energy cascade through the eddy sizes.

4.2 Turbulence modelling

Turbulence modelling ranges from weather forecasts to virtual prototyping of novel cars, airplanes, heat exchangers, gas-turbine engines, chemical reactors etc. Accurate simulations of turbulent flows are therefore of high interest both for society and in industry. Hence, turbulence modelling is one of the key elements in CFD. This section

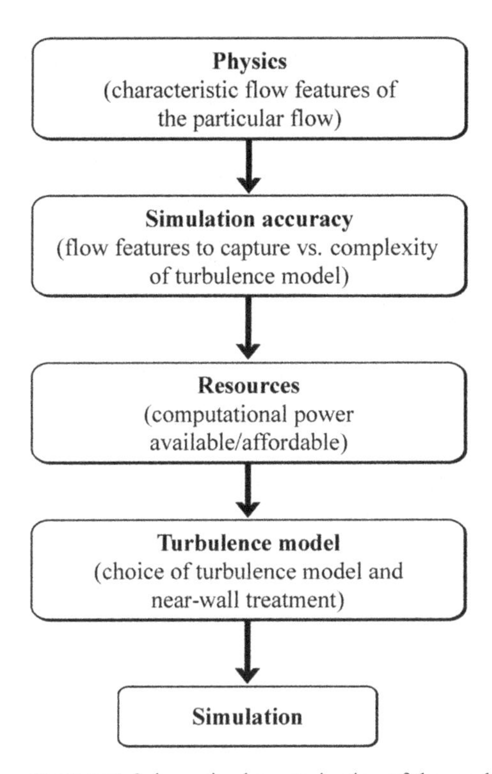

Figure 4.10 Schematic characterization of the modelling process.

will discuss some of the numerous approaches to modelling turbulent flows that have been suggested in the literature.

Unfortunately turbulent flows are characterized by fluctuating velocity fields in which there exist small-scale and high-frequency fluctuations. Thus an enormous amount of information is required if one is to describe turbulent flows completely. High-Reynolds-number flows are therefore too computationally expensive to simulate in detail. Fortunately, we usually require something less than the complete time history of every flow property over all spatial coordinates. Instead of simulating the exact governing equations, these equations can be manipulated to remove the small-scale high-frequency fluctuations, resulting in a modified set of equations that it is computationally less expensive to solve. As a consequence of the manipulation, the modified equations contain additional unknown variables. Hence, turbulence models are needed in order to determine these variables. Turbulence modelling can therefore be described as the process of closing the modified Navier–Stokes equations by providing required turbulence models.

During the last few decades numerous turbulence models of varying complexity have been proposed. The selection among these models is crucial for a successful simulation. In the ideal case the selection process will be a straightforward procedure as shown in Figure 4.10. As indicated in Figure 4.10, knowledge about the flow (i.e. whether or not the flow involves separation, whether the features of the flow result from anisotropy etc.) significantly simplifies the decision by reducing the number of turbulence models

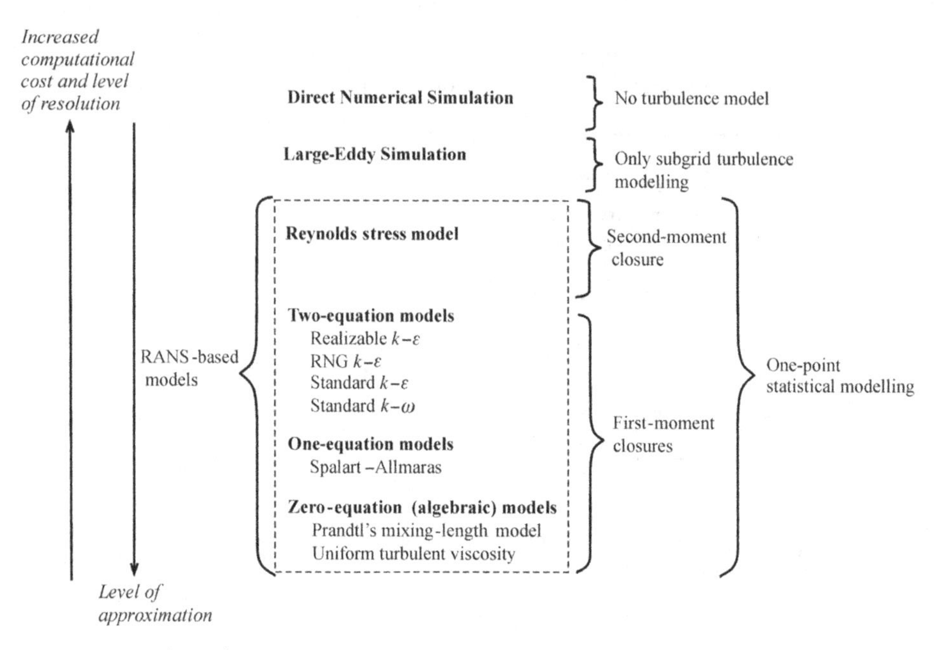

Figure 4.11 A schematic overview of turbulence modelling.

that can be used. Furthermore, in practical engineering applications the selection of turbulence models can be restricted by the computational resources that are available or affordable. It is not always the case that the required simulation accuracy implies the use of a turbulence model that can be matched with the available computational power.

In this section we give an overview of the most commonly used turbulence models and their limitations. Be forewarned that no models exist for general purposes and that every model must be used with care. A general trend for the turbulence models is that the fewer approximations are introduced, the more computational power is required. Figure 4.11 gives an overview of the different turbulence models discussed in this section. As is clearly illustrated in the figure, there is a trade-off between model accuracy and computational cost.

The choice among the models presented in Figure 4.11 is crucial for successful simulations. For simple flows good predictions can be obtained with simple turbulence models such as one-equation models. Even though the result may be less accurate for complex flows, such models may still allow one to screen effects of various design changes early in a project. The quality of the simulations is then reduced to obtaining information about trends rather than obtaining an overall accurate prediction. With the rapid development of computers and CFD codes it is expected that routine simulations will incorporate increasingly more advanced turbulence models in the future. Nevertheless, during the next few decades simulations of engineering applications will be based on turbulence models with various levels of approximation.

4.2.1 Direct numerical simulation

Direct numerical simulation (DNS) of turbulent flows may at first glance seem to be the most obvious and straightforward approach with which to simulate turbulent flows. Using DNS, in which the unsteady 3D Navier–Stokes equations are solved directly, there is no need for a turbulence model since the equations correctly describe fluid flows both for laminar and for turbulent conditions. The difficulty is actually solving these equations at high Reynolds number. Recall that a wide range of length scales and timescales exists in high-Reynolds-number turbulent flows. This means that all these turbulent scales must be resolved in the simulations. The scale of resolution needed is roughly described by the Kolmogorov length scale and timescale. Hence, very dense computational grids and short time steps are required. Add also that the equations are nonlinear and it is clear that this will challenge the computational solution algorithms, thus making any simulation very time-consuming. In fact, it will not be possible to perform DNS for real engineering problems until many more generations of computers have come and gone. Even if it were possible to perform these simulations for practical engineering applications, the amount of data would be overwhelming. At the present time DNS is a research tool rather than an aid to engineering design. The computational cost of DNS is high and it increases as the cube of the Reynolds number. For the interested reader, it can be mentioned that the computational cost for DNS of reactive flows is given by

$$t \propto Re^3 Sc^2. \tag{4.25}$$

Note that for gases $Sc \sim 1$, for liquids like water $Sc \sim 10^3$ and for very viscous liquids $Sc \sim 10^6$. DNS is thus mainly used for gaseous flows at moderate Reynolds numbers due to the high computational cost. Such deterministic simulations are useful for developing closures for statistical turbulence models and for validation of these models, but in practical engineering simulations DNS is less useful. Since DNS is of no use for practical engineering flow simulations, it will not be discussed further in this chapter.

4.2.2 Large-eddy simulation

Since the main problem in simulating high-Reynolds-number flows is the presence of very small length scales and timescales, a logical solution is to filter the equations, thus resolving only intermediate-to-large turbulence scales. Large-eddy simulation (LES) extends the usefulness of DNS for practical engineering applications by intentionally leaving the smallest turbulence scales unresolved. In LES the dynamics of the large scales are computed explicitly. Therefore, LES needs to be 3D and transient. The advantages of LES arise from the fact that the large eddies, which are hard to model in a universal way since they are anisotropic, are simulated directly. In contrast, small eddies are more easily modelled since they are closer to isotropy and adapt quickly to maintain a dynamic equilibrium with the energy-transfer rate imposed by the large eddies. The effects of the non-resolved scales, which cannot be neglected, are accounted for with subgrid stress models. This means that the subgrid models should be universal. In LES the turbulence scales are usually resolved down to the inertial subrange, Figure 4.9. As a general rule of thumb at least 80% of the turbulence energy should be resolved in the calculated velocities. Substantial savings in computational cost are realized in LES, since a coarser grid is employed than that required for DNS, Figure 4.11. This allows modelling of flows at higher Reynolds number. Nevertheless, the computational cost for LES is high in comparison with those of other turbulence models. The high computational cost stems from a very fine grid, short time steps and long computational time taken to obtain reliable statistics.

Filtering of the Navier–Stokes equations

The governing equations for LES are obtained by spatially filtering over small scales. A generalized filter can be defined by

$$\overline{U_i}(x,t) = \iiint G(x-\xi;\,\Delta)U_i(\xi,t)\mathrm{d}^3\xi, \tag{4.26}$$

where the filter function is interpreted as acting to keep values of U_i occurring on scales larger than the filter width Δ. Basically the filter function, G, is some function that is effectively zero for values of U_i occurring at the small scales. Thus, filtering eliminates eddies whose scales are smaller than the filter width. Examples of such filters are the box, Gaussian and sharp spectral filters etc. [11]. By filtering the Navier–Stokes equations, the scales which will be modelled are separated from those which will be calculated directly.

The velocity field has the decomposition,

$$U_i(x,t) = \overline{U_i}(x,t) + u_i(x,t). \tag{4.27}$$

An important difference between this decomposition and the Reynolds decomposition, which is widely used in turbulence modelling, is that $\overline{U}_i$ is a random field and that in general the filtered residual is not zero,

$$\overline{u_i}(x,t) \neq 0. \tag{4.28}$$

The filtered continuity equation is

$$\frac{\partial \overline{U_j}}{\partial x_j} = 0 \tag{4.29}$$

and the filtered momentum equation is

$$\frac{\partial \overline{U}_i}{\partial t} + \frac{\partial \overline{U}_i\overline{U}_j}{\partial x_j} = -\frac{1}{\rho}\frac{\partial \overline{P}}{\partial x_j} + \nu\frac{\partial^2 \overline{U}_i}{\partial x_j\,\partial x_j} - \frac{\partial \tau_{ij}}{\partial x_j}. \tag{4.30}$$

The closure problem in LES arises from the residual stress tensor, τ_{ij}, which is also commonly referred to as the subgrid stress tensor. Here the term τ_{ij} describes the transfer of momentum by turbulence at scales that are smaller than the filter,

$$\tau_{ij} = \overline{U_iU_j} - \overline{U}_i\overline{U}_j. \tag{4.31}$$

The filtered velocities $\overline{U}_i$ and $\overline{U}_j$ are solved for but the correlation $\overline{U_iU_j}$ is unknown and a satisfactory subgrid stress model must be provided for τ_{ij}. The simplest models usually involve modelling of a subgrid viscosity. An important difference between the subgrid viscosity and the traditional eddy viscosity is that it acts as a correction to the behaviour of the small scales, not as a correction of the entire influence of turbulence on the mean flow. This means that the subgrid viscosity is small compared with the eddy viscosity used in the eddy-viscosity models.

One frequently used model is the Smagorinsky–Lilly model. In this model the isotropic stress is included in the modified filtered pressure

$$\tilde{p} = \overline{p} + \frac{2}{3}\tau_{ii} \tag{4.32}$$

and the anisotropic part is modelled using a linear eddy-viscosity model,

$$\tau_{ij} - \tfrac{1}{3}\tau_{kk}\delta_{ij} = -2\mu_t \overline{S}_{ij}. \tag{4.33}$$

The subgrid viscosity is then calculated from

$$\mu_\mathrm{t} = \rho L_S^2 \left|\overline{S}\right|, \tag{4.34}$$

where

$$\left|\overline{S}\right| = \sqrt{2\overline{S}_{ij}\overline{S}_{ij}} \qquad \text{and} \qquad L_S = \min(\kappa d, C_\mathrm{S} V^{1/3}), \tag{4.35}$$

κ is the von Kármán constant, d is the distance to the nearest wall and V is the volume of the computational cell. The Smagorinsky coefficient C_S is of the order of 0.17 but depends, unfortunately, on the flow conditions and grid size. The interested reader is referred to [7, 11] for more information about LES. More advanced LES models, e.g. dynamic LES, also estimate the Smagorinsky coefficient by filtering on two different scales.

4.2.3 Reynolds decomposition

Industrial applications of LES are expected to increase in the near future. However, like DNS, LES is currently too computationally expensive for routine simulations. In many cases the industrial and academic communities need even simpler models than LES. In this chapter we introduce turbulence models that are widely used for simulations of engineering applications. These models are based on a method by which the scales can be separated. Recall that in the LES approach the small length scales and timescales were filtered out. However, even the intermediate-to-large turbulence scales must be filtered out to obtain a set of equations that can be used for routine simulations. Hence, the solution of these equations remains the only viable means for routine simulations of turbulent flows encountered in engineering practice. More than 100 years ago Reynolds proposed that the instantaneous variables could be split into a mean part and a fluctuating part,

$$U_i = \langle U_i \rangle + u_i \tag{4.36}$$

and

$$P = \langle P \rangle + p. \tag{4.37}$$

This method is therefore referred to as Reynolds decomposition. All turbulence models in the following sections of this chapter share the fact that they are mathematically based on the Reynolds-decomposition concept. With Reynolds decomposition, the flow is described statistically by the mean flow velocity and the turbulence quantities. By time averaging over a reasonable time period the turbulence fluctuations are separated from the non-turbulence quantities. Hence, the set of equations obtained with this method is called the Reynolds averaged Navier–Stokes equations, or simply the RANS equations. In many practical cases it is necessary to simulate non-steady flows, where the instantaneous variables are averaged over a time period that is large compared with the turbulence timescales but small compared with the timescale of the mean components. This means that the time derivative of the mean flow in the RANS equations accounts for variations at timescales larger than those of turbulence.

Recall that for incompressible flows the continuity equation reads

$$\frac{\partial U_j}{\partial x_j} = 0 \tag{4.38}$$

and the Navier–Stokes equations read

$$\frac{\partial U_i}{\partial t} + U_j \frac{\partial U_i}{\partial x_j} = -\frac{1}{\rho}\frac{\partial P}{\partial x_i} + \nu \frac{\partial^2 U_i}{\partial x_j \, \partial x_j}. \tag{4.39}$$

The equations for the mean variables of these quantities are derived by substituting the decomposed form into the Navier–Stokes equations and taking the average. Let us now substitute for the instantaneous variables in Eqs. (4.38) and (4.39) the decomposed variables. By substituting we obtain

$$\frac{\partial(\langle U_i \rangle + u_i)}{\partial x_i} = 0 \tag{4.40}$$

and

$$\frac{\partial(\langle U_i \rangle + u_i)}{\partial t} + (\langle U_j \rangle + u_j)\frac{\partial(\langle U_i \rangle + u_i)}{\partial x_j} = -\frac{1}{\rho}\frac{\partial(\langle P \rangle + p)}{\partial x_i} + \nu \frac{\partial^2(\langle U_i \rangle + u_i)}{\partial x_j \partial x_j}. \tag{4.41}$$

After decomposing the dependent variables into mean and fluctuating quantities we then time-average the equations using the operator

$$\langle \phi \rangle = \frac{1}{\tau} \int_t^{t+\tau} \phi(x, \tilde{t}) \mathrm{d}\tilde{t}. \tag{4.42}$$

Equation (4.40) then reads

$$\left\langle \frac{\partial(\langle U_i \rangle + u_i)}{\partial x_i} \right\rangle = 0 \tag{4.43}$$

and Eq. (4.41) becomes

$$\left\langle \frac{\partial(\langle U_i\rangle + u_i)}{\partial t} \right\rangle + \left\langle (\langle U_j\rangle + u_j)\, \frac{\partial(\langle U_i\rangle + u_i)}{\partial x_j} \right\rangle$$
$$= -\frac{1}{\rho} \left\langle \frac{\partial(\langle P\rangle + p)}{\partial x_i} \right\rangle + \nu \left\langle \frac{\partial^2(\langle U_i\rangle + u_i)}{\partial x_j\, \partial x_j} \right\rangle. \quad (4.44)$$

Note that all terms linear in fluctuating variables give zero on averaging:

$$\langle u_i\rangle = \langle u_j\rangle = 0. \quad (4.45)$$

This does not apply for the nonlinear term:

$$\langle U_i U_j\rangle = \langle (\langle U_i\rangle + u_i)\,(\langle U_j\rangle + u_j)\rangle = \langle (\langle U_i\rangle\,\langle U_j\rangle + u_j\langle U_i\rangle + u_i\langle U_j\rangle + u_i u_j)\rangle$$
$$= \langle U_i\rangle\langle U_j\rangle + \langle u_j\rangle\langle U_i\rangle + \langle u_i\rangle\langle U_j\rangle + \langle u_i u_j\rangle = \langle U_i\rangle\langle U_j\rangle + \langle u_i u_j\rangle\,. \quad (4.46)$$

Equations (4.43) and (4.44) reduce to

$$\frac{\partial\langle U_i\rangle}{\partial x_i} = 0 \quad (4.47)$$

and

$$\frac{\partial\langle U_i\rangle}{\partial t} + \langle U_j\rangle\,\frac{\partial\langle U_i\rangle}{\partial x_j} + \left\langle \frac{\partial u_i u_j}{\partial x_j} \right\rangle = -\frac{1}{\rho}\frac{\partial\langle P\rangle}{\partial x_i} + \nu\frac{\partial^2\langle U_i\rangle}{\partial x_j\,\partial x_j}. \quad (4.48)$$

By re-arranging we obtain the general form of the RANS equation

$$\frac{\partial\langle U_i\rangle}{\partial t} + \langle U_j\rangle\,\frac{\partial\langle U_i\rangle}{\partial x_j} = -\frac{1}{\rho}\frac{\partial\langle P\rangle}{\partial x_i} + \nu\frac{\partial^2\langle U_i\rangle}{\partial x_j^2} - \frac{\partial\langle u_i u_j\rangle}{\partial x_j}, \quad (4.49)$$

which can be written as

$$\frac{\partial\langle U_i\rangle}{\partial t} + \langle U_j\rangle\,\frac{\partial\langle U_i\rangle}{\partial x_j} = -\frac{1}{\rho}\frac{\partial}{\partial x_j}\left\{\langle P\rangle\delta_{ij} + \mu\left(\frac{\partial\langle U_i\rangle}{\partial x_j} + \frac{\partial\langle U_j\rangle}{\partial x_i}\right) - \rho\langle u_i u_j\rangle\right\}. \quad (4.50)$$

This equation was first derived by Reynolds in 1895 and is very similar to the original Navier–Stokes equations (4.39) apart from the additional term $-\rho\langle u_i u_j\rangle$. This term is referred to as the Reynolds stresses and is very important since it introduces a coupling between the mean and fluctuating parts of the velocity field. Note that sometimes $\langle u_i u_j\rangle$ is referred to as the Reynold stresses even though the precise definition of the Reynolds stresses is $-\rho\langle u_i u_j\rangle$. Since the Reynolds stress term contains products of the velocity fluctuations this term must be modelled in order to close Eq. (4.49). This is the sole purpose of RANS turbulence modelling.

Note that by averaging over all timescales of turbulence one performs averaging over the longest timescales of turbulence. This means that the only dynamic behaviour that will be resolved in the simulations is that of the mean flow. For flows that are statistically steady the Reynolds averaged equations of motion do not resolve any of the dynamics

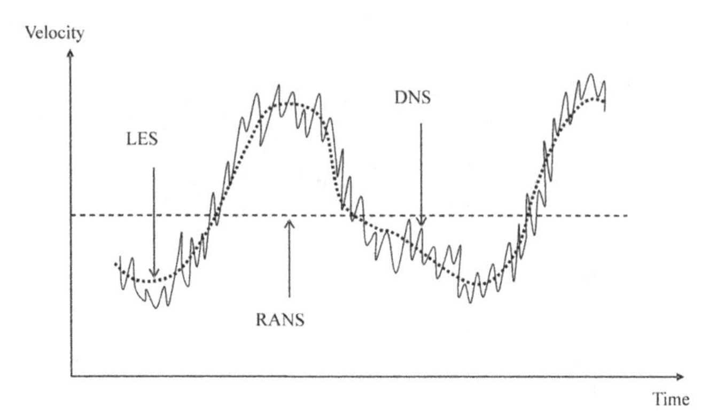

Figure 4.12 One-point representation of resolved turbulence scales in a steady turbulent flow.

even though the effects of turbulence on mass-, momentum- and heat-transfer rates are accounted for. Figure 4.12 illustrates that DNS, LES and RANS resolve scales of different sizes. Note that, while DNS resolves all scales, LES resolves only the largest scales. The non-resolved scales in LES which are more universal can be seen as small-scale fluctuations superposed on the large scales.

Reynolds stresses

Recall that the statistical-averaging process introduced unknown correlations into the mean-flow equations, namely the Reynolds stresses, $\tau_{ij} = -\rho\langle u_i u_j\rangle$. The Reynolds stress term is a second-order tensor that represents a second-order moment of the velocity components at a single point in space. These stresses appear as an additional fictitious stress tensor in Eq. (4.49) by which the fluctuating parts interact and force the mean flow. Therefore they are often called apparent stresses. It is important to point out that, although the Reynolds stress term formally appears similar to the viscous stress term, it is not part of the fluid stress but instead represents the average momentum flux due to the velocity fluctuations, thus characterizing the transfer of momentum by turbulence.

The individual Reynolds stresses in the stress tensor, $\tau_{ij} = -\rho\langle u_i u_j\rangle$, are

$$\tau_{ij} = \begin{bmatrix} -\rho\langle u_1 u_1\rangle & -\rho\langle u_1 u_2\rangle & -\rho\langle u_1 u_3\rangle \\ -\rho\langle u_2 u_1\rangle & -\rho\langle u_2 u_2\rangle & -\rho\langle u_2 u_3\rangle \\ -\rho\langle u_3 u_1\rangle & -\rho\langle u_3 u_2\rangle & -\rho\langle u_3 u_3\rangle \end{bmatrix}. \tag{4.51}$$

Since the Reynolds stress tensor is symmetric, $\langle u_1 u_2\rangle = \langle u_2 u_1\rangle$, $\langle u_1 u_3\rangle = \langle u_3 u_1\rangle$ and $\langle u_2 u_3\rangle = \langle u_3 u_2\rangle$ there are three normal stresses $-\rho\langle u_1 u_1\rangle$, $-\rho\langle u_2 u_2\rangle$ and $-\rho\langle u_3 u_3\rangle$ and three shear stresses $-\rho\langle u_1 u_2\rangle$, $-\rho\langle u_1 u_3\rangle$ and $-\rho\langle u_2 u_3\rangle$. This means that the Reynolds stress tensor contains six unknown terms that must be modelled. One straightforward approach to close Eq. (4.49) would be to derive transport equations for the Reynolds stresses. Unfortunately, this results in third-order moments of the velocity components. An attempt to derive equations describing the evolution of the third-order moments yields equations containing fourth-order moments. This goes on indefinitely and is referred to as the closure problem.

In high-Reynolds-number flows, the Reynolds stress tensor can easily be 1000 times larger than the mean viscous stress tensor,

$$\rho \left\| \langle u_i u_j \rangle \right\| \gg \mu \left\| \frac{\partial \langle U_i \rangle}{\partial x_j} + \frac{\partial \langle U_j \rangle}{\partial x_i} \right\|. \tag{4.52}$$

This means that at high Reynolds numbers the viscous stress may generally be neglected when compared with the Reynolds stress insofar as the mean flow equations are concerned. However, this does not apply towards boundaries of the flow. In these regions (thin layers adjacent to the walls) the turbulence is damped, the mean velocity gradients grow larger and the viscous stress becomes comparable to the Reynolds stress. Thus, in this region we have

$$\rho \left\| \langle u_i u_j \rangle \right\| \approx \mu \left\| \frac{\partial \langle U_i \rangle}{\partial x_j} + \frac{\partial \langle U_j \rangle}{\partial x_i} \right\|. \tag{4.53}$$

A discussion about these viscous regions and how they can be modelled is given in Section 4.3.

The Boussinesq approximation

To summarize, when the RANS equations were derived we introduced the Reynolds stresses, which are unknown terms, without adding any extra equations. This means that the Reynolds stresses must be modelled since the total number of unknowns is more than the total number of equations. Unless some assumptions are made about the Reynolds stresses it is not possible to solve Eqs. (4.47) and (4.49). One way to model the Reynolds stresses is to relate them to the dependent variables they are meant to transport. Such turbulence models make the prediction of turbulence feasible with a reasonable amount of computer time.

A simple approximation is to express the Reynolds stress tensor, $-\rho\langle u_i u_j \rangle$, in terms of the mean velocity itself. This closure has an approximate character, which means that the solution of the RANS equations is always an approximation. In what follows we consider closures that are based on the concept of turbulent eddy viscosity.

The Boussinesq approximation is based on the assumption that the components of the Reynolds stress tensor are proportional to the mean velocity gradients. The Boussinesq relation proposes that the transport of momentum by turbulence is a diffusive process and that the Reynolds stresses can be modelled using a turbulent viscosity (eddy viscosity), which is analogous to molecular viscosity. The Boussinesq approximation reads

$$\frac{\tau_{ij}}{\rho} = -\langle u_i u_j \rangle = \nu_{\mathrm{T}} \left(\frac{\partial \langle U_i \rangle}{\partial x_j} + \frac{\partial \langle U_j \rangle}{\partial x_i} \right) - \frac{2}{3} k \delta_{ij}, \tag{4.54}$$

or

$$\frac{\tau_{ij}}{\rho} = -\langle u_i u_j \rangle = \nu_{\mathrm{T}} S_{ij} - \frac{2}{3} k \delta_{ij}, \tag{4.55}$$

where

$$S_{ij} = \frac{1}{2}\left(\frac{\partial\langle U_i\rangle}{\partial x_j} + \frac{\partial\langle U_j\rangle}{\partial x_i}\right)$$

is the strain-rate tensor and k is the turbulent kinetic energy per unit mass. The turbulent kinetic energy per unit mass is defined as half the trace of the Reynolds stress tensor $k = \frac{1}{2}\langle u_i u_i\rangle$. In contrast to the molecular viscosity, the turbulent viscosity is not a fluid property but depends strongly on the state of turbulence.

The second term on the right-hand side of Eq. (4.54) represents normal stresses. Thus, a term analogous to the pressure occurs in the usual stress tensor for a viscous fluid, which can be absorbed into the real pressure term. Hence we finally obtain

$$\frac{\partial\langle U_i\rangle}{\partial t} + \langle U_j\rangle\frac{\partial\langle U_i\rangle}{\partial x_j} = -\frac{1}{\rho}\frac{\partial\left(\langle P\rangle + \frac{2}{3}\rho k\right)}{\partial x_i} + \frac{\partial}{\partial x_j}\left[(\nu + \nu_{\mathrm{T}})\frac{\partial\langle U_i\rangle}{\partial x_j}\right], \tag{4.56}$$

or

$$\frac{\partial\langle U_i\rangle}{\partial t} + \langle U_j\rangle\frac{\partial\langle U_i\rangle}{\partial x_j} = -\frac{1}{\rho}\frac{\partial\langle P\rangle}{\partial x_i} - \frac{2}{3}\frac{\partial k}{\partial x_i} + \frac{\partial}{\partial x_j}\left[(\nu + \nu_{\mathrm{T}})\left(\frac{\partial\langle U_i\rangle}{\partial x_j} + \frac{\partial\langle U_j\rangle}{\partial x_i}\right)\right]. \tag{4.57}$$

Thus, if specific details of the turbulence are not important we can interpret the fluid itself as a pseudo-fluid with an increased viscosity (effective viscosity, $\nu_{\mathrm{eff}} = \nu + \nu_{\mathrm{T}}$) that roughly approximates the turbulent mixing processes to diffusion of momentum and other flow properties. This means that diffusion models convection, where $\partial\langle u_i u_j\rangle/\partial x_j$ is transport of $\langle U_i\rangle$ in Eq. (4.49).

Regardless of the approach used to determine ν_{T}, there are several limitations with the Boussinesq approximation. Among others, the Boussinesq approximation assumes that eddies behave like molecules, that turbulence is isotropic and that there exists local equilibrium between stress and strain. As a consequence of these assumptions, predictions of simple flows may fail, e.g. in channel flows, where measurements show that $-\langle u_1^2\rangle \neq -\langle u_2^2\rangle \neq -\langle u_3^2\rangle$. Hence, models that are based on the Boussinesq approximation are limited to prediction of isotropic flows in local equilibrium. Despite the shortcomings of the Boussinesq approximation it is one of the cornerstones in several turbulence models. The reasons for this are the cost of using more elaborate turbulence models and problems with obtaining closures for higher moments.

4.2.4 Models based on the turbulent viscosity hypothesis

Most models evaluating the Reynolds stress tensor simplify the situation through the Boussinesq eddy-viscosity concept. Thus, as part of quantitative turbulence modelling this eddy viscosity, ν_{T}, must be determined. So when it comes to turbulence modelling based on the RANS equations and the eddy-viscosity concept, the turbulence model can be seen as the set of equations that are needed to determine this viscosity.

As in the kinetic theory of gases, the viscosity is proportional to velocity times distance. The turbulent-viscosity models are based on appropriate velocity, u, and length scales, l, describing the local turbulent viscosity, ν_T. The dimension of ν_T is [m^2 s^{-1}], which means that the product of these two scales gives the right dimension; that is, we have

$$\nu_T = C_\nu \frac{l^2}{t} = C_\nu ul. \tag{4.58}$$

In this expression u and l are the characteristic scales for the large turbulent eddies and C_ν is a proportionality constant. This is reasonable because these scales are responsible for most of the turbulent transport. The turbulent eddy viscosity, ν_T, in Eq. (4.57) may vary with position and time and must be specified before the set of equations can be closed. A proper turbulence model therefore involves a closed set of equations, i.e. the total number of unknown variables equals the total number of equations. There exist numerous methods that provide values for the turbulent viscosity. These methods are usually categorized by the number of additional transport equations that are required for closure, i.e. the equations required to determine the velocity and length scales describing the local turbulence. In the simplest categories of models no additional transport equations are used. These models are therefore referred to as zero-equation models. Furthermore, there are one- and two-equation models in which respectively one and two PDEs are solved together with the RANS equations. The degree of accuracy of each turbulence model depends on the validity of the assumptions behind it. In what follows we will discuss some of the zero-, one- and two-equation models that are used for this purpose. Among these models, the two-equation models are the most widely used for routine simulation of engineering applications.

Zero-, one- and two-equation models

The number of additional PDEs considered in addition to the RANS and continuity equations is used to classify the turbulence models. Hence algebraic models are classified as zero-equation models. One of the most well-known zero-equation models is Prandtl's mixing-length model. In this model ν_T is calculated using an analogy between the chaotic motions of eddies and the random motion of molecules in gases (kinetic gas theory). The mixing length, l, depends strongly on the nature of the flow and is generally space-dependent. This model offers an improvement over the constant-viscosity models and is capable of predicting some simple flows. The motivation for developing more advanced models than the zero-equation models is that it is very difficult to estimate the distribution of the mixing length. Another limitation with this model is that the eddy viscosity is instantaneously affected by the shear rate and vanishes whenever the velocity gradient is zero. This does not agree with experimental data. The reason for this is that the turbulent stresses result not only from events at a single point but also from events in the region, since they are transported by convection and diffusion and they have a certain lifetime before the energy is dissipated. Hence, zero-equation models are unsuitable for general use since they do not account for effects of accumulation and transport on turbulence.

Since turbulent eddies have a certain lifetime and are transported by convection, turbulence is not completely determined by the local conditions but depends also on the history of the eddies. One way to overcome the limitations of the zero-equation models is to relate the turbulent viscosity to a transported turbulent quantity instead of relating it to the mean velocity gradient. In contrast to the zero-equation models the one-equation models allow history effects to be accounted for. It was independently suggested by Kolmogorov and Prandtl that the square root of the time-averaged turbulent kinetic energy, k, should be employed as the characteristic turbulent velocity scale, u. Examples of such models include Prandtl's k–l model and the Spalart–Allmaras model. From the above discussion it can be concluded that transport of the turbulent kinetic energy is taken into account in one-equation models at the cost of solving one additional PDE. The problem with one-equation models is that only the characteristic velocity scale is determined from a transport equation, and the length scale must therefore be specified algebraically. An obvious solution would then be to determine the length scale from an additional transport equation. That is actually what is done in two-equation models.

Two-equation models

Zero- and one-equation models are still used for certain applications but they are rarely used for general-purpose flow simulations. For general-purpose flow simulations, the more sophisticated two-equation models are frequently used. As the number of equations increases, the computational cost increases as well. It should be mentioned that the two-equation models are sometimes referred to as complete models, since they allow the turbulent velocity and length scales to be determined independently. For practical engineering purposes, the most successful models involve two or more transport equations. This is due to the fact that it takes two quantities to characterize the length and velocity scales of turbulent flows. Using transport equations to describe these variables means that the turbulence-production and -dissipation processes can have localized rates. Without the transport mechanism turbulence has to instantly adjust to local conditions, thereby giving unrealistically large production and dissipation rates. In many cases a high local value of turbulence is due to convection of upstream-generated eddies.

A straightforward approach to model the turbulent velocity and length scales is to solve the k equation for the velocity scale and the l equation for the length scale. Usually the names of the turbulence models are logical in the sense that they reflect what is modelled. Hence, a two-equation model that models k and l is simply called a k–l model. More often the second transport equation describes transport of some property other than the length scale, l. Obviously it must be possible to determine the length scale explicitly from this property. Generally we can write an arbitrary property, ϕ, which is related to the length scale, l, in the following way:

$$\phi = k^{\alpha} l^{\beta}. \tag{4.59}$$

Table 4.2 Commonly used properties for determination of the turbulence length scale

α	β	ϕ	Alternative symbol to ϕ	Interpretation of ϕ
0	1	l	l	Length scale
1	−2	k/l^2	ω	Vorticity scale
1/2	−1	$k^{1/2}/l$	f	Frequency scale
−1/2	1	$k^{-1/2}l$	τ	Timescale
3/2	−1	$k^{3/2}/l$	ε	Dissipation rate

This means that for the k–l model $\alpha = 0$ and $\beta = 1$, hence $\phi = l$. There are many possible choices for the second turbulence variable, resulting in different values of α and β. Some of the proposed variables are given in Table 4.2.

The energy-dissipation rate, ε, is the most commonly used of these variables. As implied by its name, the k–ε model describes turbulence using two variables, namely the turbulent kinetic energy, k, and the energy-dissipation rate, ε. The relation between the turbulence length scale and the energy-dissipation rate is

$$\phi = k^{3/2}/l = \varepsilon. \tag{4.60}$$

The length scale is simply the turbulent velocity, $\sqrt{k}$, times the lifetime of the turbulent eddies k/ε,

$$l = \sqrt{k}\frac{k}{\varepsilon} = \frac{k^{3/2}}{\varepsilon}, \tag{4.61}$$

and the turbulent viscosity is given by

$$\nu_{\mathrm{T}} = C_\nu ul = C_\nu k^{1/2}\frac{k^{3/2}}{\varepsilon} = C_\nu\frac{k^2}{\varepsilon}. \tag{4.62}$$

Two-equation models are widely used for simulation of engineering applications. Even though these models impose limitations, they continue to be favourable since they are robust and inexpensive to implement.

The standard k–ε model

The k–ε model has become very popular due to the important role played by ε in the interpretation of turbulence in addition to the fact that ε appears directly in the transport equation for k. This turbulence model provides a good compromise between generality and economy for many CFD problems. Before looking at the modelled transport equations for k and ε, one should be aware that these equations are actually simplifications of the exact transport equations for k and ε. This means that the k–ε model is one of several possible closures by which the RANS equations are simplified even further. In the following section we will look particularly at the closures introduced to solve the exact k and ε equations. Note that these types of closures are not unique to the k–ε model but are in fact required to close all models that are based on statistical averaging, due the higher-order moments that are introduced. The exact transport equation for the turbulent kinetic energy, k, can be deduced from the equation for the kinetic energy Eq. (2.26) by Reynolds decomposition and reads

$$\underbrace{\frac{\partial k}{\partial t}}_{\mathrm{I}} + \underbrace{\langle U_j \rangle \frac{\partial k}{\partial x_j}}_{\mathrm{II}} = \underbrace{-\langle u_i u_j \rangle \frac{\partial \langle U_i \rangle}{\partial x_j}}_{\mathrm{III}} \underbrace{- \nu \left\langle \frac{\partial u_i}{\partial x_j} \frac{\partial u_i}{\partial x_j} \right\rangle}_{\mathrm{IV}} + \frac{\partial}{\partial x_j}\left(\underbrace{\nu \frac{\partial k}{\partial x_j}}_{\mathrm{V}} - \underbrace{\frac{\langle u_i u_i u_j \rangle}{2}}_{\mathrm{VI}} - \underbrace{\frac{\langle u_j p \rangle}{\rho}}_{\mathrm{VII}} \right). \tag{4.63}$$

The physical interpretation of the terms in the Eq. (4.63) is as follows.

I. Accumulation of k.
II. Convection of k by the mean velocity.
III. Production of k, large eddies extract energy from the mean flow.
IV. Dissipation of k by viscous stress, whereby turbulent kinetic energy is transformed into heat.
V. Molecular diffusion of k.
VI. Turbulent transport by velocity fluctuations.
VII. Turbulent transport by pressure fluctuations.

In Eq. (4.63) the terms III, IV, VI and VII are unknown and, unless some approximations are introduced, it is not possible to solve this equation. Hence, closures are required for the production, dissipation and diffusion terms. The production term represents the production of turbulent kinetic energy due to the mean flow strain rate. If the equation for the kinetic energy of the mean flow is considered, this term actually appears as a sink in the equation. This clearly shows that the production of turbulent kinetic energy is indeed a result of the mean flow losing kinetic energy, recalling that $\langle E \rangle = \overline{E} + k$. Note that the production term is the Reynolds stresses times the shear rates and maximum production will occur where both are large. This is mainly in the boundary layers close to the walls, and for flow parallel to the wall the maximum production is at $y^+ \approx 12$. The Reynolds stresses can be identified in the production term and it is assumed that the Boussinesq approximation can be used to model this term by relating it to gradients of the mean flow

$$-\langle u_i u_j \rangle = \nu_{\mathrm{T}} \left(\frac{\partial \langle U_i \rangle}{\partial x_j} + \frac{\partial \langle U_j \rangle}{\partial x_i} \right) - \frac{2}{3} k \delta_{ij}. \tag{4.64}$$

This means that the production of turbulent kinetic energy can be modelled as

$$-\langle u_i u_j \rangle \frac{\partial \langle U_i \rangle}{\partial x_j} = \nu_{\mathrm{T}} \left(\frac{\partial \langle U_i \rangle}{\partial x_j} + \frac{\partial \langle U_j \rangle}{\partial x_i} \right) \frac{\partial \langle U_i \rangle}{\partial x_j} - \frac{2}{3} k \frac{\partial \langle U_i \rangle}{\partial x_i}. \tag{4.65}$$

Note that the last term in Eq. (4.65) is zero for incompressible flow due to continuity. It is important to recall that the Boussinesq approximation is an isotropic model for the Reynolds stresses and assumes that the normal stresses are all equal.

The second closure needed to model the k equation is a relation for the energy-dissipation rate, which is the rate of destruction of turbulent kinetic energy. Note that the energy dissipation of turbulent kinetic energy is defined as

$$\varepsilon = \nu \left\langle \frac{\partial u_i}{\partial x_j} \frac{\partial u_i}{\partial x_j} \right\rangle. \tag{4.66}$$

The third closure is required to describe the turbulent transport of k. These higher-order moments (terms VI and VII) are usually modelled by assuming a gradient-diffusion transport mechanism. This assumption allows the turbulent transport due to velocity and pressure fluctuations to be modelled as

$$-\frac{\langle u_i u_i u_j \rangle}{2} - \frac{\langle u_j p \rangle}{\rho} = \frac{\nu_\mathrm{T}}{\sigma_k} \frac{\partial k}{\partial x_j}. \tag{4.67}$$

Here, σ_k is a model coefficient known as the Prandtl–Schmidt number and ν_T is the turbulent viscosity. Substituting these closures into the exact transport equation for k gives the modelled equation for k

$$\frac{\partial k}{\partial t} + \langle U_j \rangle \frac{\partial k}{\partial x_j} = \nu_\mathrm{T} \left[\left(\frac{\partial \langle U_i \rangle}{\partial x_j} + \frac{\partial \langle U_j \rangle}{\partial x_i} \right) \frac{\partial \langle U_i \rangle}{\partial x_j} \right] - \varepsilon + \frac{\partial}{\partial x_j} \left[\left(\nu + \frac{\nu_\mathrm{T}}{\sigma_k} \right) \frac{\partial k}{\partial x_j} \right]. \tag{4.68}$$

To close the k equation we need to calculate ε and the turbulent viscosity. Obviously, the energy-dissipation rate is modelled with a second transport equation. Note that the production of turbulent kinetic energy is modelled as the product of the turbulent viscosity and average velocity gradients.

The exact ε equation can be written as

$$\begin{aligned}
&\underbrace{\frac{\partial \varepsilon}{\partial t}}_{\mathrm{I}} + \underbrace{\langle U_j \rangle \frac{\partial \varepsilon}{\partial x_j}}_{\mathrm{II}} \\
&= \underbrace{-2\nu \left(\left\langle \frac{\partial u_i}{\partial x_k} \frac{\partial u_j}{\partial x_k} \right\rangle + \left\langle \frac{\partial u_k}{\partial x_i} \frac{\partial u_k}{\partial x_j} \right\rangle \right) \frac{\partial \langle U_i \rangle}{\partial x_j}}_{\mathrm{III}} \underbrace{- 2\nu \left\langle u_k \frac{\partial u_i}{\partial x_j} \right\rangle \frac{\partial^2 \langle U_i \rangle}{\partial x_k \partial x_j}}_{\mathrm{IV}} \\
&\quad \underbrace{- 2\nu \left\langle \frac{\partial u_i}{\partial x_k} \frac{\partial u_i}{\partial x_j} \frac{\partial u_k}{\partial x_j} \right\rangle}_{\mathrm{V}} \underbrace{- 2\nu\nu \left\langle \frac{\partial^2 u_i}{\partial x_k \partial x_j} \frac{\partial^2 u_i}{\partial x_k \partial x_j} \right\rangle}_{\mathrm{VI}} \\
&\quad + \frac{\partial}{\partial x_j} \left(\underbrace{\nu \frac{\partial \varepsilon}{\partial x_j}}_{\mathrm{VII}} \underbrace{- \nu \left\langle u_j \frac{\partial u_i}{\partial x_j} \frac{\partial u_i}{\partial x_j} \right\rangle}_{\mathrm{VIII}} \underbrace{- 2 \frac{\nu}{\rho} \left\langle \frac{\partial p}{\partial x_j} \frac{\partial u_j}{\partial x_j} \right\rangle}_{\mathrm{IX}} \right).
\end{aligned} \tag{4.69}$$

The physical interpretation of some of the terms in Eq. (4.69) is as follows.

I. Accumulation of ε.
II. Convection of ε by the mean velocity.
III and IV. Production of dissipation, due to interactions between the mean flow and the products of the turbulent fluctuations.
V and VI. Destruction rate of the dissipation, due to turbulent velocity fluctuations.
VII. Viscous diffusion of ε.
VIII. Turbulent transport of ε due to velocity fluctuations.
IX. Turbulent transport of ε due to pressure–velocity fluctuations.

In Eq. (4.69) there are several unknown terms containing correlations of fluctuating velocities and gradients of fluctuating velocities and pressure, namely terms III, IV, V, VI, VIII and IX. Again we need several closures for the unknown terms in order to close the equation. The result of introducing closures is that the modelled equation is drastically simplified. To avoid tedious manipulation of the exact ε equation we simply give the general form of the modelled ε equation:

$$\underbrace{\frac{\partial \varepsilon}{\partial t}}_{\text{I}} + \underbrace{\langle U_j \rangle \frac{\partial \varepsilon}{\partial x_j}}_{\text{II}} = \underbrace{C_{\varepsilon 1} \nu_{\text{T}} \frac{\varepsilon}{k} \left[\left(\frac{\partial \langle U_i \rangle}{\partial x_j} + \frac{\partial \langle U_j \rangle}{\partial x_i} \right) \frac{\partial \langle U_i \rangle}{\partial x_j} \right]}_{\text{III}} - \underbrace{C_{\varepsilon 2} \frac{\varepsilon^2}{k}}_{\text{IV}} + \underbrace{\frac{\partial}{\partial x_j} \left[\left(\nu + \frac{\nu_{\text{T}}}{\sigma_\varepsilon} \right) \frac{\partial \varepsilon}{\partial x_j} \right]}_{\text{V}}. \tag{4.70}$$

The physical interpretation of the terms in Eq. (4.70) is as follows.

I. Accumulation of ε.
II. Convection of ε by the mean velocity.
III. Production of ε.
IV. Dissipation of ε.
V. Diffusion of ε.

The time constant for turbulence is calculated from the turbulent kinetic energy and the rate of dissipation of turbulent kinetic energy:

$$\tau = k/\varepsilon. \tag{4.71}$$

Note that the source term in the ε equation is the same as that in the k equation divided by the time constant τ in Eq. (4.71) and the rate of dissipation of ε is proportional to

$$\varepsilon/\tau = \varepsilon^2/k. \tag{4.72}$$

The turbulent viscosity must be calculated to close the k–ε model. Recall that the turbulent viscosity is given as the product of the characteristic velocity and length scales, $\nu_{\text{T}} \propto ul$. This means that we have

$$\nu_{\text{T}} = C_\mu \frac{k^2}{\varepsilon}. \tag{4.73}$$

Table 4.3 Closure coefficients in the standard k–ε model

Constant	Value
C_μ	0.09
$C_{\varepsilon 1}$	1.44
$C_{\varepsilon 2}$	1.92
σ_k	1.00
σ_ε	1.30

Finally, the five closure coefficients (C_μ, $C_{\varepsilon 1}$, $C_{\varepsilon 2}$, σ_k and σ_ε) in the k–ε model are assumed to be universal and thus constant, although they can vary slightly from one flow to another. The values for these constants are given in Table 4.3.

The robustness and the easily interpreted model terms make the k–ε model the most widely used two-equation model. However, the standard k–ε model does not always give good accuracy. Examples of flows that cannot be predicted accurately with the standard k–ε model are flows with streamline curvature, swirling flows and axisymmetric jets. The inaccuracies stem from the underlying Boussinesq hypothesis which imposes isotropy and from the way in which the dissipation equation is modelled. Actually this model was derived and tuned for flows with high Reynolds numbers. This implies that it is suited for flows in which the turbulence is nearly isotropic and flows in which the energy cascade proceeds in local equilibrium with respect to generation. Furthermore, the model parameters in the k–ε model are a compromise to give the best performance for a wide range of different flows. The accuracy of the model can therefore be improved by adjusting the parameters for particular experiments. As the strength and weaknesses of the standard k–ε model have become known, improvements have been made to the model to improve its performance. In the literature, numerous modifications for the turbulence models have been suggested. The most well-known variants of the standard model are the RNG and the realizable k–ε models. It is not within the scope of this book to discuss all these modifications; however, a closer look will be taken at the RNG and realizable k–ε models. In fact, the k–ε model and its variants have become a workhorse in practical engineering flow simulations.

The RNG k–ε model

The main physical difference between the standard model and the RNG k–ε model lies in a different formulation of the dissipation equation. In the RNG k–ε model, an additional source term, S_ε, is added and the equation is given by

$$\frac{\partial \varepsilon}{\partial t} + \langle U_j \rangle \frac{\partial \varepsilon}{\partial x_j} = C_{\varepsilon 1} \nu_{\mathrm{T}} \frac{\varepsilon}{k} \left[\left(\frac{\partial \langle U_i \rangle}{\partial x_j} + \frac{\partial \langle U_j \rangle}{\partial x_i} \right) \frac{\partial \langle U_i \rangle}{\partial x_j} \right] - C_{\varepsilon 2} \frac{\varepsilon^2}{k} + \frac{\partial}{\partial x_j} \left[\left(\nu + \frac{\nu_{\mathrm{T}}}{\sigma_\varepsilon} \right) \frac{\partial \varepsilon}{\partial x_j} \right] - S_\varepsilon, \tag{4.74}$$

where the source term S_ε is given by

$$S_\varepsilon = 2\nu S_{ij} \left\langle \frac{\partial u_l}{\partial x_i} \frac{\partial u_l}{\partial x_j} \right\rangle. \tag{4.75}$$

In the RNG k–ε model the additional term, S_ε, is modelled as

$$S_\varepsilon = \frac{C_\mu \eta^3 (1 - \eta/\eta_0)\, \varepsilon^2}{\left(1 + \beta\eta^3\right) k}. \tag{4.76}$$

Here

$$\eta = \frac{k}{\varepsilon}\sqrt{2S_{ij}S_{ij}}$$

and S_{ij} is the strain-rate tensor. The constants η_0 and β take the values 4.38 and 0.012, respectively. This additional term is an ad-hoc model that is largely responsible for the differences in performance compared with the standard model.

The standard k–ε model is known to be too dissipative, namely the turbulent viscosity in recirculations tends to be too high, thus damping out vortices. In regions with large strain rate the additional term in the RNG model results in smaller destruction of ε, hence augmenting ε and reducing k, which in effect reduces the effective viscosity. Improvements can therefore be expected for swirling flows and flows in which the geometry has a strong curvature. Hence the RNG model is more responsive to the effects of rapid strain and streamline curvature than the standard k–ε model. Although the RNG model is very good for predicting swirling flows, its predictions for jets and plumes are inferior to the standard k–ε model. By using a mathematical renormalization-group (RNG) technique, the k–ε model can be derived from the Navier–Stokes equation, which results in different, analytical, model constants. The constants stemming from the RNG analysis differ slightly from the empirically determined constants in the standard k–ε model.

The realizable k–ε model

The realizable model differs from the standard k–ε model in that it features a realizability constraint on the predicted stress tensor, thereby giving it the name of realizable k–ε model. The difference comes from a correction of the k equation where the normal stress can become negative in the standard k–ε model for flows with large mean strain rates. This can be seen by analysing the normal components of the Reynolds stress tensor:

$$\langle u_i u_i \rangle = \sum_i \left\langle u_i^2 \right\rangle = \frac{2}{3}k - 2\nu_{\mathrm{T}} \frac{\partial \langle U_i \rangle}{\partial x_j}. \tag{4.77}$$

Note that the normal stress $\langle u_i u_i \rangle$ must be larger than zero by definition, since it is a sum of squares. However, Eq. (4.77) implies that, if the strain is sufficiently large, the normal stress becomes negative. The realizable k–ε model uses a variable C_μ so that this will never occur. In fact, C_μ is no longer taken to be a constant; instead, it is a function of the local state of the flow to ensure that the normal stresses are positive under all flow conditions, i.e. to ensure realizability. In other words, the realizable k–ε model ensures that the normal stresses are positive, i.e. $\langle u_i^2 \rangle \geq 0$. Neither the standard nor the RNG k–ε model is realizable. Realizability also means that the stress tensor satisfies $\langle u_i^2 \rangle \langle u_j^2 \rangle - \langle u_i u_j \rangle^2 \geq 0$, i.e. Schwarz's inequality is fulfilled. Hence, the model is likely to provide better performance for flows involving rotation and separation.

In addition, this model generally involves a modification of the ε equation. This modification involves a production term for turbulent energy dissipation that is not found in either the standard or the RNG model. The standard k–ε model predicts the spreading rate in planar jets reasonably well, but poorly predicts the spreading rate for axisymmetric jets. This is considered to be mainly due to the modelled dissipation equation. It is noteworthy that the realizable k–ε model resolves the round-jet anomaly, i.e. it predicts the spreading rate for axisymmetric jets as well as for planar jets. It is important to realize that this model is better suited to flows in which the strain rate is large. This includes flows with strong streamline curvature and rotation. Validation of complex flows, e.g. boundary-layer flows, separated flows and rotating shear flows, has shown that the realizable k–ε model performs better than the standard k–ε model.

The k–ω models

Another popular two-equation model is the k–ω model. In this turbulence model the specific dissipation, ω, is used as the length-determining quantity. This quantity is called specific dissipation by definition, where $\omega \propto \varepsilon/k$, and it should be interpreted as the inverse of the timescale on which dissipation occurs. The modelled k equation is

$$\frac{\partial k}{\partial t} + \langle U_j \rangle \frac{\partial k}{\partial x_j} = \nu_{\mathrm{T}} \left[\left(\frac{\partial \langle U_i \rangle}{\partial x_j} + \frac{\partial \langle U_j \rangle}{\partial x_i} \right) \frac{\partial \langle U_i \rangle}{\partial x_j} \right] - \beta k \omega + \frac{\partial}{\partial x_j} \left[\left(\nu + \frac{\nu_{\mathrm{T}}}{\sigma_k} \right) \frac{\partial k}{\partial x_j} \right] \tag{4.78}$$

and the modelled ω equation is

$$\begin{aligned} \frac{\partial \omega}{\partial t} + \langle U_j \rangle \frac{\partial \omega}{\partial x_j} = {} & \alpha \frac{\omega}{k} \nu_{\mathrm{T}} \left[\left(\frac{\partial \langle U_i \rangle}{\partial x_j} + \frac{\partial \langle U_j \rangle}{\partial x_i} \right) \frac{\partial \langle U_i \rangle}{\partial x_j} \right] - \beta^* \omega^2 \\ & + \frac{\partial}{\partial x_j} \left[\left(\nu + \frac{\nu_{\mathrm{T}}}{\sigma_\omega} \right) \frac{\partial \omega}{\partial x_j} \right], \end{aligned} \tag{4.79}$$

where the turbulent viscosity is calculated from

$$\nu_{\mathrm{T}} = \frac{k}{\omega}. \tag{4.80}$$

An advantage of this model compared with the k–ε model is the performance in regions with low turbulence when both k and ε approach zero. This causes problems because both k and ε must go to zero at a correct rate since the dissipation term in the ε equation includes ε^2/k. In contrast, no such problems occur in the k–ω model. Furthermore, the k–ω model has been shown to reliably predict the law of the wall when the model is used in the viscous sub-layer, thereby eliminating the need to use wall functions, except for computational efficiency. The k–ω model proves to be superior in this area due to the fact that the k–ε model requires either a low-Reynolds-number modification or the use of wall functions when applied to wall-bounded flows. However, the low-*Re* k–ω model requires a very fine mesh close to the wall with the first grid below $y^+ = 5$.

For constant-pressure boundary-layer flow, both the k–ε model and the k–ω model give good predictions. However, for boundary layers with adverse pressure gradients the k–ω model is claimed to give better predictions. For further information the reader is referred to [12].

4.2.5 Reynolds stress models (RSMs)

Turbulence models based on the Boussinesq approximation are inaccurate for flows with sudden changes in the mean strain rate. This is because history effects of the Reynolds stresses persist for long distances in turbulent flows due to a relatively slow exchange of momentum between eddies. Recall that the Boussinesq approximation assumes that eddies behave like molecules and exchange momentum quickly. In the Reynolds stress models, the isotropic eddy-viscosity concept, which is the primary weakness of the two equation models, is not used. Abandoning the isotropic eddy-viscosity concept, the RSM closes the RANS equations via solution of the transport equations for Reynolds stresses, $\tau_{ij} = -\rho\langle u_i u_i\rangle$, and for the energy-dissipation rate, ε. The RSM solves one equation for each Reynolds stress and hence does not need any modelling of the turbulence to the first order. The Reynolds stress models are nonlinear eddy-viscosity models, and are usually referred to as second-moment closures (second-order closures) since the only terms modelled are of third order or higher. The primary advantage of stress-transport models is the natural approach in which non-local and history effects are accounted for. These models can significantly improve the performance under certain conditions, since they account for effects of streamline curvature, swirl, rotation and rapid changes in strain rate in a more rigorous manner than do the two-equation models. In principle, stress-transport modelling is a much better approach but the problem is in providing closures to the extra unknown correlations that arise in the derivation of the exact equations. The interest in using Reynolds stress-transport equations is also held back by the fact that these equations are much more expensive to compute, since $6 + 1$ additional PDEs are solved, and they are susceptible to numerical instability since they are strongly coupled. However, the Reynolds stress model must be used when the flow features of interest are the result of anisotropy in the Reynolds stresses.

Stress-transport modelling

The stress-transport model solves one transport equation for each Reynolds stress. The equations describing the transport of Reynolds stresses can be obtained directly from the Navier–Stokes equations by using the Reynolds decomposition and averaging.

The complete transport equations for the Reynolds stresses are

$$\underbrace{\frac{\partial \langle u_i u_j \rangle}{\partial t}}_{\text{I}} + \underbrace{\langle U_k \rangle \frac{\partial \langle u_i u_j \rangle}{\partial x_k}}_{\text{II}} = \underbrace{-\left(\langle u_i u_k \rangle \frac{\partial \langle U_j \rangle}{\partial x_k} + \langle u_j u_k \rangle \frac{\partial \langle U_i \rangle}{\partial x_k} \right)}_{\text{III}}$$
$$\underbrace{-2\nu \left\langle \frac{\partial u_i}{\partial x_k} \frac{\partial u_j}{\partial x_k} \right\rangle}_{\text{IV}} + \underbrace{\left\langle \frac{p}{\rho} \left(\frac{\partial u_i}{\partial x_j} + \frac{\partial u_j}{\partial x_i} \right) \right\rangle}_{\text{V}}$$
$$\underbrace{-\frac{\partial}{\partial x_k} \left(\langle u_i u_j u_k \rangle + \delta_{ik} \frac{\langle u_j p \rangle}{\rho} + \delta_{jk} \frac{\langle u_i p \rangle}{\rho} - \nu \frac{\partial \langle u_i u_j \rangle}{\partial x_k} \right)}_{\text{VI}}. \tag{4.81}$$

The terms in Eq. (4.81) represent the following effects.

I. Accumulation of $\langle u_i u_j \rangle$.
II. Convection of $\langle u_i u_j \rangle$ by the mean velocity.
III. Production of $\langle u_i u_j \rangle$, generation rate of the turbulent stresses by mean shear, large eddies extract energy from the mean flow strain rate.
IV. Viscous dissipation of $\langle u_i u_j \rangle$, dissipation rate of turbulent stresses, whereby turbulent kinetic energy is transformed into heat.
V. Pressure–strain correlation, which gives redistribution among the Reynolds stresses.
VI. Transport terms, all terms except the last term, which is molecular diffusion, account for turbulent transport.

Equation (4.81) may be written in the following shorthand notation:

$$\frac{\partial \langle u_i u_j \rangle}{\partial t} + \langle U_k \rangle \frac{\partial \langle u_i u_j \rangle}{\partial x_k} = P_{ij} - \varepsilon_{ij} + \phi_{ij} + d_{ij}. \tag{4.82}$$

The physical interpretation of Eq. (4.82) is that the individual stresses are generated, convected and dissipated at different rates. Hence, the modelling is at a higher fundamental level than the approach of obtaining a turbulent viscosity. Consequently, the eddy-viscosity hypothesis is not needed, which eliminates one of the major shortcomings of the models described in the previous section. The main difference between the turbulent kinetic energy and stress equations is term V, which has no equivalent in the k equation. This term is called the pressure–strain or pressure-scrambling term. The term acts to redistribute turbulent energy from one stress component to another. This concept can be shown through a summation of the equations for the normal stresses by using the continuity of the diagonal elements

$$\phi_{ii} = \left\langle \frac{p}{\rho} \left(\frac{\partial u_i}{\partial x_i} + \frac{\partial u_i}{\partial x_i} \right) \right\rangle = 0. \tag{4.83}$$

This means that, if the normal stress in one direction is less than that in the other direction, it will receive energy through ϕ_{ij}. Thus, this process can be thought of as redistributive, tending to return the turbulence to an isotropic state, with no direct influence on the level of turbulence energy.

Equation (4.81) gives six equations for the Reynolds stresses. Note that terms I, II and III are exact in that they contain only the Reynolds stresses and the mean strains. Except for the first three terms, all other terms must be modelled. Actually, it is possible to derive a set of equations for the unknowns. However, this will merely introduce higher-order terms that require closure. Hence, at some level, turbulence models must be introduced in order to close the set of equations.

The dissipative terms are modelled assuming that they are isotropic since the dissipative processes occur at the smallest scales,

$$\varepsilon_{ij} = 2\nu \left\langle \frac{\partial u_i}{\partial x_k} \frac{\partial u_j}{\partial x_k} \right\rangle = \frac{2}{3} \varepsilon \delta_{ij}. \tag{4.84}$$

Here the energy dissipation, ε, is obtained from its own transport equation.

The contributions from turbulence–turbulence interactions and from mean strains are usually accounted for in modelling of the pressure–strain term as

$$\phi_{ij} = \left\langle \frac{p}{\rho} \left(\frac{\partial u_i}{\partial x_j} + \frac{\partial u_j}{\partial x_i} \right) \right\rangle = \phi_{ij1} + \phi_{ij2}. \tag{4.85}$$

Here, ϕ_{ij1} is the slow pressure–strain term which serves to redistribute energy among the Reynolds stresses and ϕ_{ij2} is the rapid pressure–strain term which counteracts the production of anisotropy. The term giving a return to isotropy, or slow pressure–strain term, ϕ_{ij1}, is modelled as

$$\phi_{ij1} = -c_1 \varepsilon \left(\frac{\langle u_i u_j \rangle}{k} - \frac{2}{3} \delta_{ij} \right) \tag{4.86}$$

and the rapid pressure–strain term, ϕ_{ij2}, is modelled as

$$\phi_{ij2} = -c_2 \left(P_{ij} - \frac{2}{3} P_{kk} \delta_{ij} \right), \tag{4.87}$$

where

$$P_{ij} = -\left(\langle u_i u_k \rangle \frac{\partial \langle U_j \rangle}{\partial x_k} + \langle u_j u_k \rangle \frac{\partial \langle U_i \rangle}{\partial x_k} \right). \tag{4.88}$$

The turbulent transport terms are often modelled on the basis of a gradient-diffusion hypothesis

$$\langle u_k \phi \rangle \propto -\frac{k}{\varepsilon} \langle u_k u_l \rangle \frac{\partial \phi}{\partial x_l} \tag{4.89}$$

and assuming a negligible pressure-diffusion. Hence, the term is modelled as

$$\begin{aligned} d_{ij} &= -\frac{\partial}{\partial x_k} \left(\langle u_i u_j u_k \rangle + \delta_{ik} \frac{\langle u_j p \rangle}{\rho} + \delta_{jk} \frac{\langle u_i p \rangle}{\rho} - \nu \frac{\partial \langle u_i u_j \rangle}{\partial x_k} \right) \\ &= \frac{\partial}{\partial x_k} \left(c_S \frac{k}{\varepsilon} \langle u_k u_l \rangle \frac{\partial \langle u_i u_j \rangle}{\partial x_l} \right). \end{aligned} \tag{4.90}$$

Note that it is good practice to include wall correction terms in the modelled equations to account for wall reflection effects. Note also that an additional transport equation for the energy-dissipation rate is solved. Thus, we have a solvable set of equations, e.g. 11 equations for 11 unknowns, $\langle U_i \rangle$, $\langle u_i u_j \rangle$, $\langle P \rangle$ and ε.

4.2.6 Advanced turbulence modelling

Turbulence models that fall beyond the bounds of the categories presented earlier have been developed. It is difficult to sum up all the progress in this field in a few sentences; instead, the interested reader is referred to textbooks on advanced turbulence modelling for further insight into this exciting area. It should also be mentioned that research in this area still remains active.

4.2.7 Comparisons of various turbulence models

In the previous sections, we have presented turbulence models that are commonly encountered in commercial CFD codes. In these sections the physical and mathematical principles underlying the turbulence models were presented together with discussions about their limitations. Table 4.4 gives a short summary of the advantages and shortcomings of these models.

In general the turbulence models are developed to predict velocities accurately. The parameters in the models, e.g. k and ε, may very well be off by a factor of 3. Using these parameters in other models, e.g. for mixing or bubble break-up, should be done with the awareness that the parameters do not have exact physical relevance but only show the trend.

4.3 Near-wall modelling

Most flows of engineering interest involve situations in which the flow is constrained by a solid wall. Ludwig Prandtl was the first to realize that the relative magnitude of the inertial and viscous forces changed on going from a layer near the wall to a region far from the wall. In the early 1900s he presented the theory which describes boundary-layer effects. The wall no-slip condition ensures that, over some region of the wall layer, viscous effects on the transport processes must be large. Recall that particular turbulence models such as the k–ε model are not valid in the near-wall region, where viscous effects are dominant. Furthermore, rapid variation of the flow variables occurs within this region. This implies that a very fine computational mesh is required in order to resolve the steep gradients of the flow variables accurately. Representation of these processes within a CFD model raises problems. Basically, there are two approaches that can be used to model the near-wall region. In the first approach the viscosity-affected near-wall region is not resolved. Instead, wall functions are used to obtain boundary conditions for the mean velocity components and the turbulent quantities at the first

Table 4.4 General advantages and disadvantages for different classes of turbulence models

Turbulence model	Advantages	Shortcomings/limitations
Zero-equation models, e.g. mixing-length model	Cost-effective model applicable for a limited number of flows.	When $\partial U/\partial y = 0 \Rightarrow \nu_T = 0$. Lack of transport of turbulent scales. Estimation of the mixing length is difficult. Cannot be used as a general turbulence model.
One-equation models, e.g. k-algebraic model	Cost-effective model applicable for a limited number of flows.	The use of an algebraic equation for the length scale is too restrictive. Transport of the length scale is not accounted for.
Two-equation models, k–ε group standard, RNG, realizable, k–ω and SST	Complete models in the sense that velocity and length scales of turbulence are predicted with transport equations. Good results for many engineering applications. Especially good for trend analysis. Robust, economical and easy to apply.	Limited to an eddy-viscosity assumption. Turbulent viscosity is assumed to be isotropic. Convection and diffusion of the shear stresses are neglected.
Standard k–ε	The most widely used and validated model.	Not good for round jets and flows involving significant curvature, swirl, sudden acceleration, separation and low-*Re* regions.
RNG k–ε	Modification of the standard k–ε model gives improved simulations for swirling flows and flow separation.	Not as stable as the standard k–ε model. Not suited for round jets.
Realizable k–ε	Modification of the standard k–ε model gives improved simulations for swirling flows and flow separation. Can also handle round jets.	Not as stable as the standard k–ε model.
k–ω model	Works well at low *Re*. Does not need wall functions. Works well with adverse pressure gradients and separating flow.	Needs fine mesh close to the wall, first grid point at $y^+ < 5$.
SST model	Uses k–ε in the free stream and k–ω in the wall-bounded region. Works well with adverse pressure gradients and separating flow. Many authors recommend that the SST model should replace the k–ε model as the first choice.	Needs fine mesh close to the wall. Overpredicts turbulence in regions with large normal strain, e.g. stagnation regions and regions with strong acceleration, but is better than k–ε.
Reynolds stress models, (RSMs)	Applicable for complex flow where the turbulent-viscosity models fail. Accounts for anisotropy. Good performance for many complex flows, e.g. swirl, flow separation and planar jets.	Computationally expensive with 11 transport equations. Several terms in the transport equations must be closed. Poor performance for some flows due to the closures introduced in the model.

Table 4.4 *(cont.)*

Turbulence model	Advantages	Shortcomings/limitations
Large-eddy Simulation (LES)	Applicable to complex flows. Gives information about structures in turbulent flows. Gives a lot of information that cannot be obtained otherwise.	High computational cost. Large amount of data that must be stored and post-processed. Difficult to find proper time-resolved boundary conditions for flow.
DNS	No turbulence models are introduced. Useful at low *Re* numbers, especially for gaseous flows. Useful to develop and validate turbulence models.	Extreme computational cost for practical engineering flow simulations. Huge amount of data.

grid point far from the wall. The other approach involves modification of the turbulence models, which allows the viscosity-affected region to be resolved.

4.3.1 Turbulent boundary layers

At a solid wall, the relative velocity between the fluid and the wall is zero. This is called the 'no-slip condition'. The relative velocity is zero because molecules, moving with random motions plus the mean fluid velocity, hit the solid wall and all the relative momentum is lost, being transferred to the solid wall. Molecules bouncing back into the flow slow down the fluid in the wall layer. Thus, a 'boundary layer' is created. In this region the velocity increases rapidly from zero at the wall to the free-stream velocity. Note that boundary layers may be either laminar or turbulent, depending on the Reynolds number. Turbulent boundary layers, from high Reynolds numbers, are characterized by unsteady swirling flows inside the boundary layer, which give higher mass-, momentum- and heat-transfer rates than apply to a laminar boundary layer, which arises from a low Reynolds number. The efficient momentum transport in turbulent boundary layers increases the wall shear stress. Thus, at high Reynolds numbers we encounter turbulent boundary layers, which produce a greater drag. Hence, details of the flow within boundary layers may be very important in many CFD simulations.

Turbulent boundary layers, of thickness δ, can be divided into an inner region, $0 < y < 0.2\delta$, and an outer region, $0.2\delta < y < \delta$, as shown in Figure 4.13.

It is common practice to divide the inner region into sub-layers on the basis of the relative magnitude of the viscous and turbulent parts of the total shear stress, τ_{xy},

$$\tau_{xy} = \rho\nu\frac{\mathrm{d}\langle U_x\rangle}{\mathrm{d}y} - \rho\langle u_x u_y\rangle. \tag{4.91}$$

In the innermost layer, the viscous sub-layer, the flow is almost laminar and molecular viscosity plays a dominant role in momentum transfer. The viscous sub-layer is defined as the region where the viscous stress is dominant. Near the wall, viscous damping reduces the tangential velocity fluctuations, while kinematic blocking reduces the normal

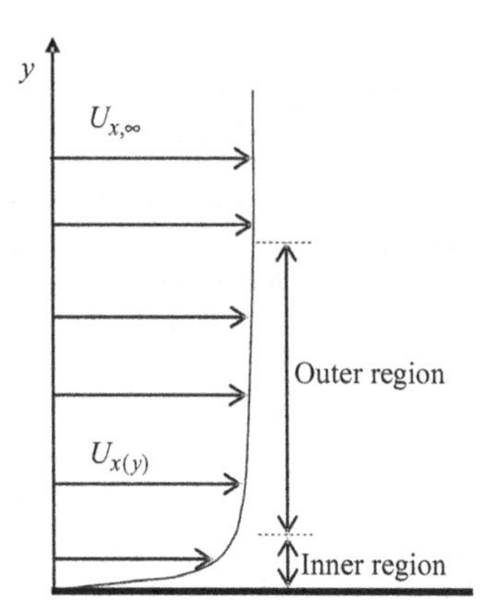

Figure 4.13 The boundary layer in a turbulent flow.

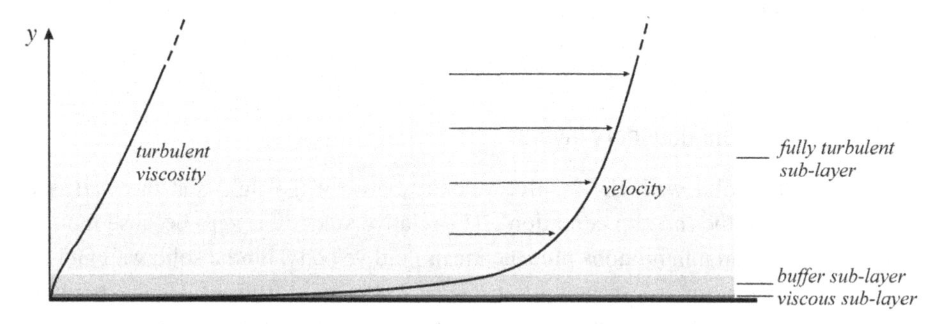

Figure 4.14 Sub-layers in the inner region.

fluctuations. Note that the boundary conditions of the velocities, u_x, $u_y \to 0$ as $y \to 0$, imply that the Reynolds stresses vanish rapidly as the wall is approached. At the wall the stress is entirely due to viscous shear,

$$\tau_w = \rho \nu \frac{\mathrm{d}\langle U_x \rangle}{\mathrm{d}y}\bigg|_{y=0}. \tag{4.92}$$

In fact, the turbulence in the boundary layer takes its origin from this region. But almost all turbulent eddies are aligned with the wall, and the effective turbulence perpendicular to the wall is almost zero in the vicinity of the wall. Further away from the wall, the viscous and turbulent stresses are equally important. This interim region is the transition layer often referred to as the buffer layer. At even larger distances from the wall the turbulent stresses become dominant. This region is called the fully turbulent layer. Here turbulence plays a major role and viscous effects are negligible. Figure 4.14 shows the three sub-layers.

It is common practice to express the physical extent of these sub-layers in terms of wall variables. The characteristic velocity scale for the sub-layers is given by

$$u_* = \sqrt{\tau_\mathrm{w}/\rho}, \tag{4.93}$$

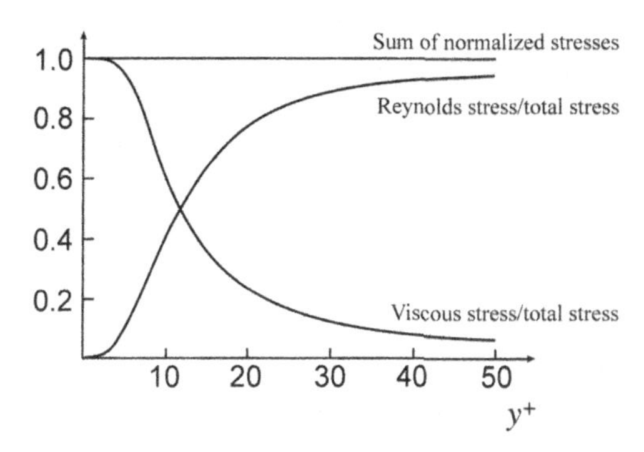

Figure 4.15 Near-wall stresses in the inner region of the turbulent boundary layer.

where u_* is the wall friction velocity, which is of the same order as the r.m.s. value of the velocity fluctuations. This allows us to define a characteristic wall length scale as

$$l_* = \nu/u_*. \tag{4.94}$$

Note that the Reynolds number based on u_* and l_* is equal to one, thus l_* determines the domain of the flow which is significantly affected by the viscosity. On the basis of the characteristic velocity and length scales it is common to use scaled variables to express the physical extent of the sub-layers:

$$u^+ = U/u_* \tag{4.95}$$

and

$$y^+ = y/l_* = yu_*/\nu. \tag{4.96}$$

The following classification of the inner region, which is based on experimental investigations, is commonly used.

(1) Viscous sub-layer $0 < y^+ < 5$.
(2) Buffer sub-layer $5 < y^+ < 30$.
(3) Fully turbulent sub-layer $30 < y^+ < 400$ ($y/\delta = 0.1$–0.2).

The viscous and turbulent stresses, given as functions of the y^+ values, are shown in Figure 4.15. Note that the total shear stress is almost constant over the inner region and is approximately equal to τ_w. Thus, this region is often referred to as the constant-stress layer.

In many situations it is inevitable that the boundary layer becomes detached from the wall. Boundary layers tend to separate from walls when there is an increasing fluid pressure in the direction of the flow; this is known as an adverse pressure gradient. While adverse pressure gradients reduce the wall shear stress through decreasing the flow velocity close to the wall, separation is often associated with a large increase in drag. Turbulent boundary layers can prevent or delay separation, thereby reduce the drag significantly. Thus, numerous methods by which to avoid or delay separation by creating turbulent boundary layers have been invented. The dimples on a golf ball are a good

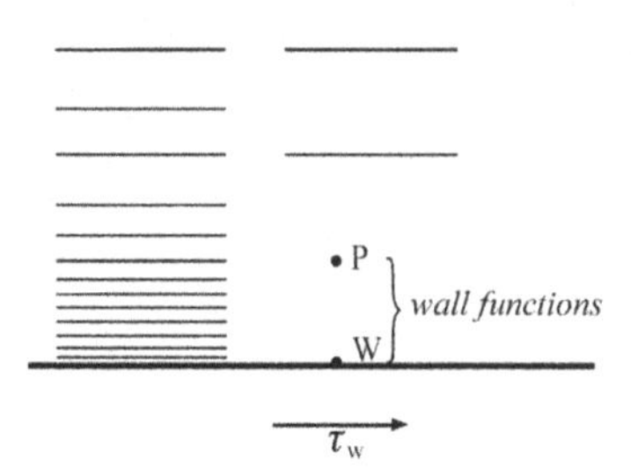

Figure 4.16 Application of wall functions to CFD simulation.

example of this. The interested reader can find more information about boundary-layer theory in [13].

4.3.2 Wall functions

Wall functions are empirical rules that are based on the logarithmic law of the wall. The wall functions may be needed in order to avoid having dense meshes in CFD simulations or they may be needed since particular turbulence models are not valid in the viscosity-affected near-wall region. The wall functions estimate the velocities $\langle U_i \rangle$, k and ε or $\langle u_i u_j \rangle$ in the RANS models in the first cell close to the wall. A wall function is also used for estimation of temperature, T, and concentration, C, in heat- and mass-transfer simulations.

Standard wall functions

The basic idea of the wall-function approach is to apply boundary conditions some distance away from the wall so that the turbulence model is not solved close to the wall. The wall functions allow calculations to be carried out with the first grid point, P, in the region where the wall function is valid, rather than on the wall itself, as shown in Figure 4.16. The boundary conditions are used at P, which represents the first grid point, and W represents the corresponding point on the wall. Thus, the wall functions allow the rapid variations of flow variables that occur within the near-wall region to be accounted for without resolving the viscous near-wall region.

In addition, the use of wall functions obviates the need to modify the turbulence models to account for the viscosity-affected near-wall regions. The mean velocity in the inner region of the boundary layer can be formulated in the universal form

$$\langle U \rangle^+ = f\left(y^+\right). \tag{4.97}$$

Assuming that the total stress τ_w is constant and the turbulent part of the total stress tensor is negligible in the viscous sub-layer, Eq. (4.91) reduces to

$$\frac{\tau_w}{\rho} = \nu \frac{d\langle U_x \rangle}{dy}. \tag{4.98}$$

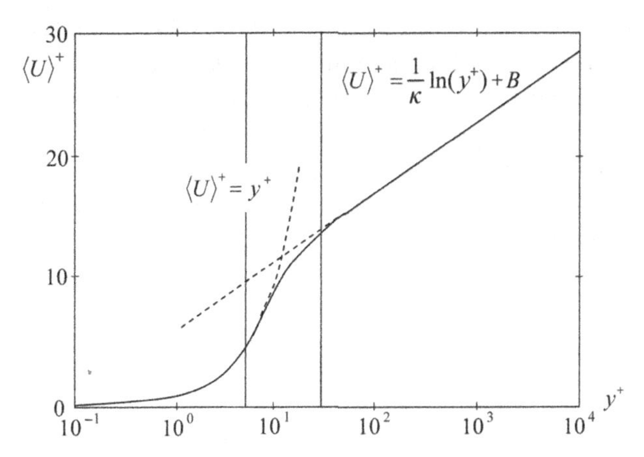

Figure 4.17 The law of the wall.

Integrating with respect to y and applying the no-slip boundary condition gives

$$\langle U_x \rangle = \frac{\tau_w y}{\rho \nu} = \frac{u_*^2 y}{\nu} \tag{4.99}$$

or in dimensionless form

$$\langle U_x \rangle^+ = y^+. \tag{4.100}$$

In the fully turbulent layer, the total stress tensor reduces to $\tau_{xy} = -\rho \langle u_x u_y \rangle$. Since the shear stress is almost constant over the inner region of the boundary layer and is approximately equal to τ_w, we obtain

$$\tau_w = -\rho \langle u_x u_y \rangle. \tag{4.101}$$

By introducing Prandtl's mixing-length model and the relation $l = \kappa y$, we obtain

$$\frac{\tau_w}{\rho} = -\langle u_x u_y \rangle = l^2 \left(\frac{d\langle U_x \rangle}{dy} \right)^2 = \kappa^2 y^2 \left(\frac{d\langle U_x \rangle}{dy} \right)^2. \tag{4.102}$$

Recall that the characteristic velocity scale for the sub-layers is given by $u_* = \sqrt{\tau_w/\rho}$. Equation (4.102) can now be written as

$$u_*^2 = \kappa^2 y^2 \left(\frac{d\langle U_x \rangle}{dy} \right)^2. \tag{4.103}$$

On taking the square root of both sides and integrating with respect to y, we obtain the logarithmic velocity profile, which in dimensionless form reads

$$\langle U_x \rangle^+ = \frac{1}{\kappa} \ln(y^+) + B, \tag{4.104}$$

where $\kappa \approx 0.42$ and $B \approx 5.0$ (κ is the von Kármán constant). Equation (4.104) is referred to as the logarithmic law of the wall or simply the log law. Thus, in the viscous sub-layer the velocity varies linearly with y^+, whereas in the buffer sub-layer it approaches the log law, as shown in Figure 4.17.

Besides the logarithmic profile for the mean velocity, the wall functions also consist of equations for the near-wall turbulence quantities. There is no transport of k to the wall, while ε often has a maximum at the wall, but the derivation of the boundary conditions for the turbulence quantities is beyond the scope of this book. Note that, in the derivation of the boundary conditions for the turbulence quantities, it is assumed that the flow is in local equilibrium, which means that production equals dissipation. The boundary condition for k is given by

$$k = \frac{u_*^2}{C_\mu^{1/2}} \tag{4.105}$$

and that for ε by

$$\varepsilon = \frac{u_*^3}{\kappa y}. \tag{4.106}$$

The use of wall functions requires that the first grid point adjacent to the wall is within the logarithmic region. Ideally, the first grid point should be placed as close to the lower bound of the log-law region as possible in order to get as many grid points in the boundary layer as possible. In terms of dimensionless distance, that is $30 < y^+ < 100$. The wall-function approach saves considerable computational resources because the viscosity-affected near-wall region does not need to be resolved. The log law has proven very effective as a universal function for the inner region of the flat-plate turbulent boundary layer and it has been verified experimentally in numerous studies. Wall functions can successfully be used in many CFD simulations, and most CFD programs adjust the wall function accordingly when $y^+ < 30$, but $y^+ > 300$ should be avoided. However, doubts can be raised about the validity of wall functions under conditions such as strong pressure gradients and separated and impinging flows. Under such conditions the quality of the predictions is likely to be compromised. This does not mean that such flows cannot be simulated, rather that standard wall functions are not an appropriate choice. In the following sections we will introduce non-equilibrium wall functions and also a near-wall modelling approach whereby the viscous sub-layer actually is resolved.

Wall functions for non-equilibrium turbulent boundary layers

Non-equilibrium turbulent boundary layers are boundary layers that have been perturbed from the normal flat-plate boundary-layer state. Recall that the log law for the normal flat-plate boundary layer was derived under the assumption of constant shear. A local equilibrium between production and dissipation was also assumed in the derivation of the turbulence quantities. Thus, difficulties arise in applying standard wall functions when the simplifying assumptions upon which the wall functions are based are not applicable. Constant shear and the local-equilibrium hypothesis are therefore the conditions that most restrict the universality of the standard wall functions. In a boundary layer experiencing an adverse pressure gradient, the fluid closest to the wall is retarded due to the pressure increase in the streamwise direction. As a result, the wall shear stress is decreased. Consequently, adverse pressure gradients alter the mean velocity profile as

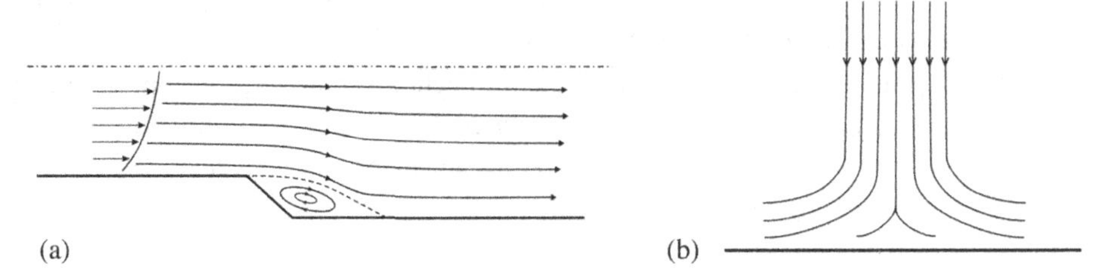

Figure 4.18 (a) Flow separation with a recirculation zone and a reattachment point. (b) Impinging flow.

well as the turbulence in the boundary layer. This means that, when the pressure gradient is strong enough, the logarithmic boundary-layer representation (the log law) cannot be used. Hence, for several flow conditions, e.g. flow separation and reattachment, strong pressure gradients and flow impinging on a wall (Figure 4.18), the flow situation departs significantly from the ideal conditions and the accuracy of the standard wall functions is low.

Modified wall functions that are capable to some extent of accounting for the effects of pressure gradients and departure from equilibrium have been developed. By using such modified wall functions for non-equilibrium boundary layers, improved predictions can be obtained. These wall functions typically consist of a log law for the mean velocity, which is sensitized to pressure-gradient effects. Boundary conditions for the turbulence quantities are derived using methods whereby the equilibrium condition is relaxed. Thus, these modifications further extend the applicability of the wall-function approach and allow improvements to be obtained for complex flow conditions.

4.3.3 Improved near-wall modelling

Improved modelling of wall-bounded flows can be achieved using a two-layer zonal approach or using low-Reynolds-number turbulence models. These techniques permit the governing equations to be solved all the way to the wall, thereby eliminating the use of wall functions and hence improving the wall shear-stress and wall heat-transfer predictions. Obviously resolution of the near-wall region including the viscous sub-layer requires a very fine near-wall grid resolution. Hence, this modelling approach requires a large amount of computational power compared with the wall-function approach.

Two-layer zonal modelling

In the two-layer zonal approach, the domain is divided into two zones or regions, as the name implies. These two regions may be identified by the wall-distance-based Reynolds number

$$Re_y = y\frac{\sqrt{k}}{\nu}, \tag{4.107}$$

where y is the distance to the nearest wall.

The fully turbulent region is normally taken to have $Re_y > 200$ and the viscosity affected region to have $Re_y < 200$. In the viscosity-affected near-wall region, a one-equation turbulence model for the turbulent kinetic energy is applied and an algebraic relation is used to determine the energy-dissipation rate. In contrast, a two-equation model such as the standard or more advanced k–ε model is employed in the fully turbulent region. Thus, in the viscous zone the energy dissipation is calculated from

$$\varepsilon = \frac{k^{3/2}}{l_\varepsilon}, \tag{4.108}$$

where l_ε is an appropriate length scale.

It is common practice to use a blending function to calculate the turbulent viscosity in the transition region. This function simply blends the turbulent viscosity in the viscosity-affected region with the turbulent viscosity in the turbulent region to obtain a smooth transition. Thus the blending function is defined as unity far from the wall and zero at the wall. The two-layer zonal approach requires approximately the same boundary-layer resolution as in the low-Reynolds-number approach. Since the energy dissipation is calculated from an algebraic equation, this approach may be more stable than the low-Reynolds-number approach.

Low-Reynolds-number turbulence models

One way of characterizing turbulence models is to distinguish between high- and low-Reynolds-number models. In the former, wall functions are used to approximate turbulence quantities close to walls. The standard k–ε model is an example of a high-Reynolds-number model. Whereas high-Reynolds-number models are valid for turbulent core flows, they are not valid in regions close to walls, where viscous effects predominate over turbulent ones. Low-Reynolds-number models are examples of models that are valid also in the viscous wall region and can thus be integrated all the way to the wall.

The low-Reynolds-number modifications typically consist of damping functions for the source terms in the transport equation for ε and in the expression for the turbulent viscosity. These modifications allow the equations to be integrated through the turbulent boundary layer, including the viscous sub-layer, thereby giving better predictions for near-wall flows. It is important to point out that these models are applicable for flows with high global Reynolds number wherein the flow is fully turbulent. These models are not useful for solving flows with low global Reynolds numbers. For that a transition model is needed. It should also be noted that these models are of ad-hoc nature and cannot be relied upon to give consistently good results for all types of flows. Low-Reynolds-number variants of the k–ε model include the Launder–Sharma and Lam–Bremhorst models.

For low-Reynolds-number models, the general transport equations for k are given by

$$\frac{\partial k}{\partial t} + \langle U_j \rangle \frac{\partial k}{\partial x_j} = \frac{\partial}{\partial x_j}\left(\left(\nu + \frac{\nu_{\mathrm{T}}}{\sigma_k}\right)\frac{\partial k}{\partial x_j}\right) + \nu_{\mathrm{T}}\left[\left(\frac{\partial \langle U_i \rangle}{\partial x_j} + \frac{\partial \langle U_j \rangle}{\partial x_i}\right)\frac{\partial \langle U_i \rangle}{\partial x_j}\right] - \varepsilon \tag{4.109}$$

and the general transport equations for ε are given by

$$\frac{\partial \tilde{\varepsilon}}{\partial t} + \langle U_j \rangle \frac{\partial \tilde{\varepsilon}}{\partial x_j} = \frac{\partial}{\partial x_j}\left(\left(\nu + \frac{\nu_T}{\sigma_\varepsilon}\right)\frac{\partial \tilde{\varepsilon}}{\partial x_j}\right) + C_{1\varepsilon} f_1 \nu_T \frac{\tilde{\varepsilon}}{k}\left[\left(\frac{\partial \langle U_i \rangle}{\partial x_j} + \frac{\partial \langle U_j \rangle}{\partial x_i}\right)\frac{\partial \langle U_i \rangle}{\partial x_j}\right]$$
$$- C_{2\varepsilon} f_2 \frac{\tilde{\varepsilon}^2}{k} + E, \tag{4.110}$$

where the turbulent viscosity is calculated from

$$\nu_T = f_\mu C_\mu \frac{k^2}{\tilde{\varepsilon}} \tag{4.111}$$

and the energy dissipation, ε, is related to $\tilde{\varepsilon}$ by

$$\varepsilon = \varepsilon_0 + \tilde{\varepsilon}. \tag{4.112}$$

The quantities ε_0 and E are defined differently for each model; ε_0 is the value of ε at the wall. The difference between these models and the standard k–ε model is the damping functions f_1 and f_2 in the transport equation of ε and the damping function f_μ. The damping functions are generally written in terms of specifically defined Reynolds numbers

$$Re_t = \frac{k^2}{\nu\varepsilon} \tag{4.113}$$

and

$$Re_y = \frac{\sqrt{k}y}{\nu}. \tag{4.114}$$

Obviously the global Reynolds number has nothing to do with the low-Reynolds-number turbulence models. The low Reynolds number comes from the local Reynolds number.

4.3.4 Comparison of three near-wall modelling approaches

As has already been pointed out, the near-wall treatment determines the accuracy of the wall stresses and of the near-wall turbulence prediction. Hence, appropriate near-wall turbulence modelling is crucial in order to capture important flow features such as flow separation, reattachment and heat- and mass-transfer rates. Figure 4.19 illustrates the implementation of the three near-wall treatment approaches mentioned in the previous sections.

The general recommendation for a standard wall function is $30 < y^+ < 100$, preferably in the lower region. At high *Re* the log law is valid up to higher y^+ and the upper limit may increase to 300–500. For low-*Re* models and enhanced wall functions the first grid point should be close to $y^+ = 1$, and there should be at least ten grid points in the viscosity-affected near-wall region, i.e. $y^+ < 20$.

The lengths corresponding to $y^+ = 1$ and $y^+ = 30$ for pipe flows at various Reynolds numbers are given in Table 4.5 to give a feeling for the actual dimensions.

Table 4.5 Physical lengths corresponding to y^+ for flow through a smooth pipe, as a function of the Reynolds number

Re	$y^+ = 1$	$y^+ = 30$
5000	145 μm	4.4 mm
50 000	20 μm	0.60 mm
500 000	2.5 μm	0.075 mm
5 000 000	0.30 μm	0.0090 mm

Table 4.6 A summary of near-wall modelling approaches

Modelling approach	Physics	Grid requirements	Numerics
Wall functions	−	+	+
Low-Reynolds-number modifications	+/−	−	+/−
Zonal modelling	+/−	−	+

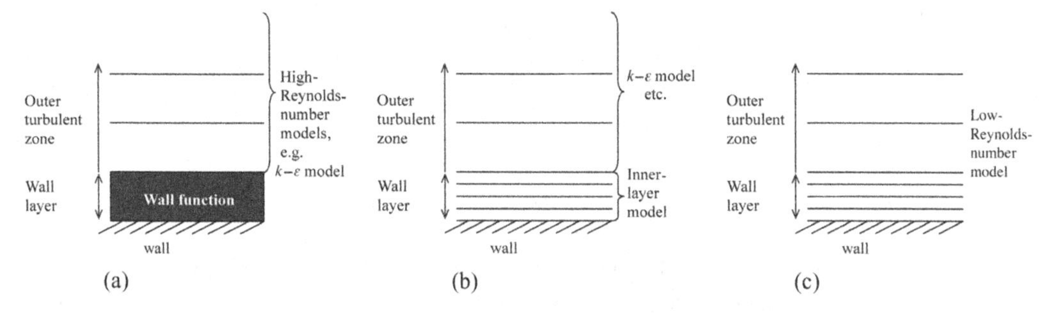

Figure 4.19 Illustrations of various near-wall treatment approaches: (a) the wall function; (b) the two-layer zonal approach; (c) low-Reynolds-number modifications.

It might not always be apparent what near-wall modelling approach to use in a certain simulation. As with the choice of selecting an appropriate turbulence model, the choice among the near-wall modelling approaches is strongly coupled to the physics of the particular flow and the computational resources available. The general pros and cons of the three approaches illustrated in Figure 4.19 are summarized in Table 4.6. This table indicates the level of physics involved, the computational power required and the numerical difficulties involved in implementing the three near-wall modelling approaches in CFD simulations.

4.4 Inlet and outlet boundary conditions

Simulation of turbulent flows requires knowledge of the turbulent quantities at all boundaries where the flow enters the computational domain. For a two-equation

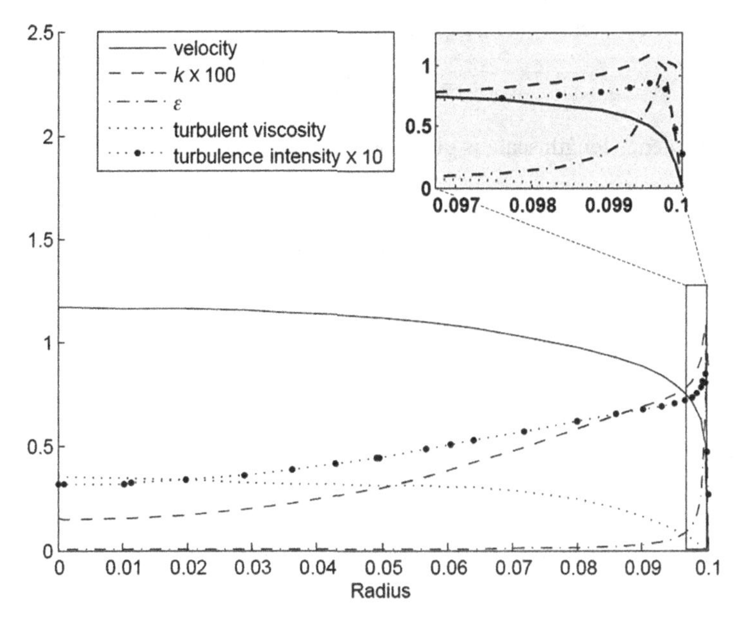

Figure 4.20 Velocity and turbulence properties in a cross-section of a pipe at $Re = 200\,000$.

turbulence model, such as the k–ε model, both the turbulent kinetic energy and the energy-dissipation rate should be specified. However, since it is the specification of the turbulence timescales, velocity scales and length scales entering the domain that defines the turbulence, it need not necessarily be the values of the turbulent kinetic energy and its dissipation rate that are specified. Instead of specifying these values, the boundary conditions are often specified in terms of the turbulence intensity, which is defined by the ratio of the fluctuating component of the velocity to the mean velocity, and a characteristic turbulence length scale. For internal flows, a turbulence intensity of 5%–10% and a length scale of 1%–10% of the hydraulic diameter are usually appropriate. Specifying boundary conditions for the Reynolds stress transport model is more difficult than for two-equation models, since all the stresses must be specified. If these are not available turbulence could be assumed to be isotropic at the inlet, i.e. zero shear stresses and the normal stresses given by $\frac{2}{3}k$. Always select boundary conditions with care since inconsistent boundary conditions may cause unrealistic reduction of the turbulence after the inlet, or turbulence may flow through the entire domain without changing.

Velocity and turbulence are usually not constant at the inlet and outlet, but depend both on upstream and on downstream conditions. The inlet and outlet should be located as far away from the region of interest as possible so that the approximations under the given conditions will not affect the results of the simulations. Figure 4.20 shows the radial variation of velocity and turbulence properties in a pipe at $Re = 200\,000$. Using the average properties for velocity, i.e. $U = 1\ \mathrm{m\ s^{-1}}$, is reasonable but k and ε have large radial variation and average values will not describe the actual inlet conditions. A better approximation is to use the turbulence intensity and turbulence length scale.

The turbulence intensity for a pipe at high Reynolds number can be estimated from

$$I = \frac{u}{\langle U \rangle} = 0.16 Re^{-1/8} \tag{4.115}$$

and the turbulence length scale is given by

$$l = 0.07L, \tag{4.116}$$

where L is the hydraulic diameter and $\overline{U}$ the average velocity. It is then possible to estimate k and ε from

$$k = \frac{3}{2}(\langle U \rangle)^2 \quad \text{and} \quad \varepsilon = C_\mu^{3/4} \frac{k^{3/2}}{l}. \tag{4.117}$$

4.5 Summary

Turbulence is of considerable importance in most flows of engineering interest, so turbulence modelling is one of the key elements to successful CFD simulations. In this chapter the physical and mathematical principles underlying turbulence modelling were explained. The purpose of this chapter was, besides providing a survey of turbulence models, also to indicate their validity and limitations in various applications. Even though we have far from exhausted all aspects, this chapter serves as an overview of turbulence modelling. Readers who are particularly interested in turbulence modelling are encouraged to turn their attention to the references given in this book for more in-depth discussions of this subject.

Questions

(1) Discuss why turbulence has to be modelled.
(2) Explain the origin of the Reynolds stresses in the RANS equations and explain what is meant by the closure problem.
(3) Explain what is meant by the Boussinesq approximation.
(4) What limitations does the Boussinesq approximation impose on a turbulence model?
(5) Discuss the application of zero-, one- and two-equation models.
(6) Discuss the differences between Reynolds stress modelling and eddy-viscosity-based modelling.
(7) Explain what is meant by large-eddy simulations.
(8) Explain what is meant by wall functions, why they are used and when it is appropriate to use them.
(9) Discuss how the near-wall treatment can be improved, i.e. when the wall-function approach is not appropriate.
(10) Discuss what turbulence boundary conditions it is appropriate to use.

5 Turbulent mixing and chemical reactions

The purpose of this chapter is to give an introduction to problem's faced by engineers wanting to use CFD for detailed modelling of turbulent reactive flows. After reading this chapter you should be able to describe the physical process of turbulent mixing and know why this can have an effect on the outcome of chemical reactions, e.g. combustion. The problem arises when the grid and time resolution is not sufficient to resolve the concentration and the average concentration in the cells is a poor estimation of the actual concentration as shown in Figure 5.1. The local concentration changes fast, and we need models that can predict the space- and time-average reaction rate in each computational cell.

The average concentration in a computational cell can be used to describe macromixing (large-scale mixing) in the reactor and is relatively straightforward to model. The concentration fluctuations, on the other hand, can be used to describe micromixing (small-scale mixing on the molecular level). To quantify micromixing, the variance of the concentration fluctuations is used. Chemical reactions can take place only at the smallest scales of the flow, after micromixing has occurred, because reactions occur only as molecules meet and interact. An expression for the instantaneous rate of chemical reactions is often known for homogeneous mixtures. However, the average rate of chemical reactions in a reactor subject to mixing will depend also on the rate of micromixing.

In modelling fast reactions it is not sufficient to know the time-average concentration since at a specific position only one of the reactants may be present for some fraction of the time and no reaction will occur. The models for the average reaction rate make use both of the average concentrations and of the variance. The time average alone will not tell us whether both reactants were present simultaneously. A large variance tells us that the instantaneous concentration is far from the average and the probability that both reactants are present is low, whereas a small variance tells us that the concentrations of both reactants are close to the average and there is a high probability that both are present simultaneously. Hence, it is of great importance to understand that the average and variance represent macromixing and micromixing on the largest and smallest scales of the flow, respectively. In this chapter we will present the tools required to describe the flow and simulate concentration variations using probability density functions (PDFs), and use them to calculate average reaction rates.

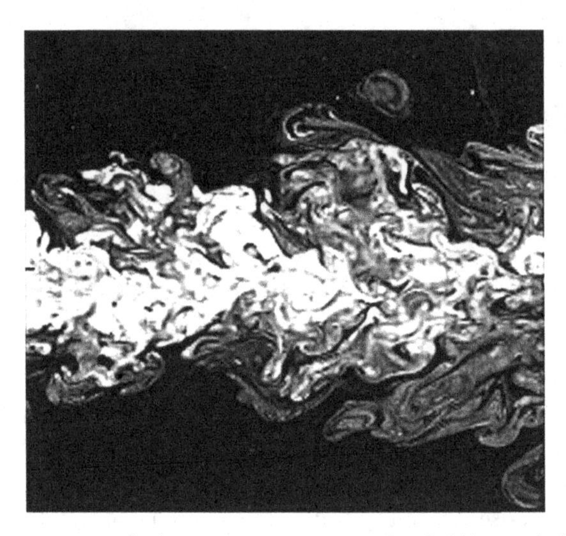

Figure 5.1 The instantaneous concentration field in a turbulent flow.

5.1 Introduction

Mathematical modelling of turbulent reactive flows is of great interest for a wide variety of applications in chemical-process engineering and combustion engineering. In chemical engineering a reactor serves, among other things, to mix species in an effort to obtain a desired reaction. It has been realized that the way in which reactants are mixed can actually be very important for the outcome of mixing-sensitive reactions. Only recently have efficient and reliable mathematical models that are able to incorporate these effects of mixing been developed. More exactly, these models are able to describe the initial mixing between fluids accurately, which can be of crucial importance. In this regard CFD is an invaluable tool since the whole reactor, including the injection, can be discretized and modelled in great detail.

Nonetheless, the process industry has been rather slow at taking advantage of this fact, and highly advanced CFD models are not very often used in the development of chemical reactors. In mechanical engineering there has always been a greater incentive for development, since there is here a major economical benefit to be derived from even the slightest improvement of a turbulent combustion engine. There are also a wider range of utilities (e.g. diesel engines, spark-ignition engines, furnaces, gas turbines), more severe reacting conditions and generally more stringent regulations for pollution, forcing improvements of existing tools. Of interest to the engineer is the fact that the bulk of the mathematical models derived for non-premixed turbulent gaseous combustion are equally valid for the liquid flows more commonly encountered in the process industry.

However, the relative success of a CFD analysis will be completely reliant on the accuracy of the mathematical model used for describing the underlying physics. Furthermore, as will be shown in this section, there will always have to be a certain level of modelling, since the physics of turbulent reactive flows is too complex to be fully resolved from first principles. In general there will have to be a trade-off between level of description and computational efficiency, in a hierarchy of models. In fact, the more of the flow is modelled, the less expensive the computations will be.

The highest-level mathematical descriptions of a turbulent reactive flow use only first principles and no modelling, i.e. direct numerical simulation (DNS). However, this approach is completely impractical for all industrial applications due to the enormous computational cost. The next-level large-eddy simulation (LES) models only the smallest scales of the flow (molecular mixing and chemical reactions), whereas the lowest level computes only statistical properties. Except for the highest level, the calculations are in general still not accurate enough to make experiments superfluous. However, for many systems there are models available that can give at least reasonably reliable predictions, completely without the aid of experiments. Still, it is our firm belief that the optimal design of a chemical reactor or combustion unit will be achieved through a balanced unification of experience, experiments and simulations.

The most common application area for reactive-mixing models is in turbulent combustion. The problem of mixing and reaction is exactly the same in combustion as in chemical reactors, but one must also take into account the sometimes large density and temperature fluctuations in combustion. This complicates the mathematical description and sometimes requires models that are used only for combustion. In combustion it is also common for the species to be premixed, whereupon reactions start only after ignition. This exact problem is not encountered in chemical reactors. In what follows we will assume that the reactants are initially separated (non-premixed), that the heat release from the reactions is small (so that the temperature fluctuations will be negligible) and that the density is constant. We will address only pure reaction-and-mixing problems, i.e. problems in which turbulent mixing has a direct effect on the outcome of chemical reactions. This is commonly referred to as the turbulence-chemistry effect. Hence we will not discuss more classical reaction problems involving more than one phase, e.g. packed beds, monoliths, reactive distillation etc.

5.2 Problem description

Danckwerts (in 1958) was the first chemical engineer to study the influence of mixing on the evolution of chemical reactions. Danckwerts established that for some reactions the way in which species were mixed could severely affect the product selectivity. This was in sharp contrast to reactor models routinely used by chemical engineers that neglected the effects of mixing, e.g. ideal batch, perfectly mixed continuously stirred tank reactors or plug flow. However, far from all reactions are mixing-sensitive, and these reactor models still serve a purpose in the process industry. In general, if chemical reactions are slow, mixing has no influence on the mean rate of reaction and the 1D ideal reactor models suffice. This follows since the reactants will be well mixed on the smallest scales (micromixing will be complete) before substantial reactions can occur. The only real problem is then that of how to obtain an accurate rate expression for the chemistry. When the typical time required for mixing is of the same order as, or longer than, the typical time required for reactions, mixing models must be introduced to describe the physics and to get realistic results for selectivity calculations. Furthermore, fast reactions depend very much on how reactants are mixed and on the geometry of the reactor. CFD simulations are then necessary for sufficient accuracy.

To illustrate the main problem in modelling of reactive flows, consider a single irreversible reaction leading to some products:

$$A + B \xrightarrow{k_1} \text{products}. \tag{5.1}$$

The chemical reaction rate is now assumed to be second order with rate constant k_1. Mathematically this can be represented as

$$S_A = S_B = -k_1 C_A C_B. \tag{5.2}$$

In statistical modelling of turbulent reactive flows one is interested in the average, not the instantaneous, rate of the chemical reactions. For this reason it is common to introduce Reynolds decomposition (see Section 4.2.3) of the instantaneous concentration into its mean and fluctuating parts. Representing the average with angle brackets and fluctuation about the average with primes, the instantaneous concentration can be decomposed as $C_\alpha = \langle C_\alpha \rangle + C'_\alpha$. Inserting for this in Eq. (5.2) and taking the average leads to the following expression for the average reaction rate:

$$\langle S_A \rangle = \langle S_B \rangle = -\langle k_1 C_A C_B \rangle = -k_1 \left(\langle C_A \rangle \langle C_B \rangle + \langle C'_A C'_B \rangle \right). \tag{5.3}$$

The time-averaged concentrations $\langle C_A \rangle$ and $\langle C_B \rangle$ are easily available in the simulations, but the covariance of the fluctuating components $\langle C'_A C'_B \rangle$ is the major problem in modelling of turbulent reactive flows. Note that for slow reactions that term will be zero, since there will be no fluctuations left when the reactions start to occur, i.e. micromixing will be complete and the mixture will be homogeneous. Hence for slow reactions there is no turbulence-chemistry effect and Eq. (5.3) will be closed. Since fast chemical reactions occur during the early stages of mixing, there will during the course of reactions be large local concentration fluctuations, as illustrated in Figure 5.2. In Figure 5.2 there are large areas containing only A or only B, where no reactions can occur. The reactions occur only when A and B are present simultaneously, and here the term $\langle C'_A C'_B \rangle$ can be significant.

The term $\langle C'_A C'_B \rangle$ is the reactive-mixing analogy to the Reynolds stress $u_i u_j$. We know from Chapter 4 that the convective Reynolds stress usually can be modelled as a diffusion process, due to the chaotic nature of turbulence. In other words, diffusion is used to model convection. Unfortunately, there is no such analogy known for $\langle C'_A C'_B \rangle$. In fact, it has been proven that in general it is not possible to model $\langle C'_A C'_B \rangle$ using merely the average (macroscale) concentrations $\langle C_A \rangle$ and $\langle C_B \rangle$, their gradients or the rate of mixing, which complicates matters significantly. In other words, it has been realized that, for sufficient accuracy in modelling of chemical reactions, greater levels of complexity are necessary than for pure flow computations. Since the complexity is so high, we will not go into great detail or give a complete review of the current state of the art. Instead, we will first discuss the nature of reactive mixing and then discuss the simplest models that can be incorporated into commercial CFD software.

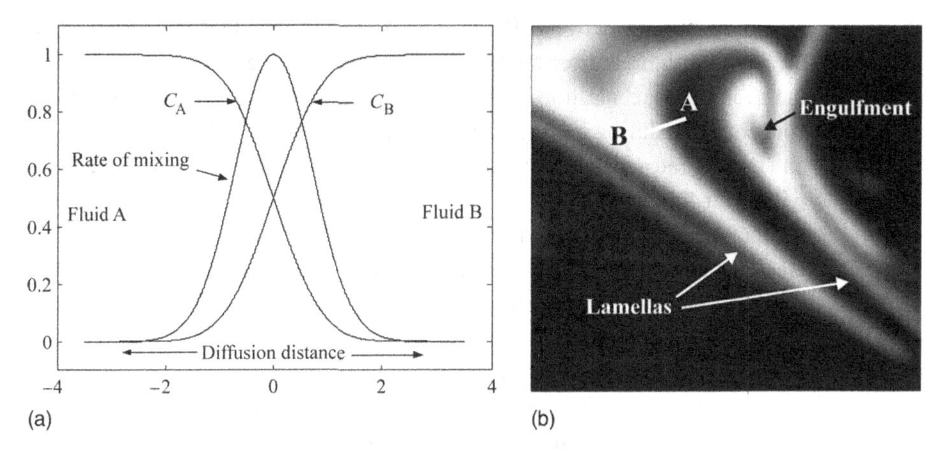

Figure 5.2 Mixing across an interface. (a) Concentration profiles across an interface between fluids A and B at the position shown with the thick white line in (b). (b) An instantaneous image of mixing between A and B in a turbulent flow field. The image is 2 mm × 2 mm. $\eta_B \approx 3$ μm and $\eta_K \approx 150$ μm (see Section 5.4), which gives a diffusion distance of about η_K.

5.3 The nature of turbulent mixing

Mixing is the process that acts on initially separated (non-premixed) fluids, resulting in a final homogeneous mixture. When white milk is poured into a cup of black coffee, the result will eventually be a new homogeneous, brown mixture. If the cup is initially stirred with a spoon the mixing time can be seconds. However, without stirring the milk can take several minutes to blend in. In other words, stirring can significantly increase the rate of mixing.

The nature of reactive mixing is quite easily understood once you realize that it is all about getting molecules to meet and interact. The most efficient way to achieve this is generally through agitation (stirred tanks) or simply by creating a high enough velocity to obtain turbulence (pipe flow). Either way, the main objective is to achieve turbulence. First of all the mean flow and the largest scales of turbulence lead to efficient macromixing, i.e. the reacting components can be quickly distributed over the whole reactor geometry. Hence, macromixing is mixing on the largest scales of the reactor. However, chemical reactions occur at the smallest scales of the flow where molecules meet and diffusion is important. In fluid mixing, turbulence is the process that acts on a fluid element, causing continuous deformation and stretching (to reduce the size of fluid elements) followed by engulfment, which occurs at the very smallest scales and significantly increases the interface area between the fluid element and the bulk. At the smallest scales the reacting components from different fluid elements diffuse into contact and thus reaction can occur.

Understanding what happens at the interfacial area between mixtures is thus of great importance. It is known from experiments that fluid elements in turbulent flows align in layers of lamellas where the concentrations across an interface can be well approximated mathematically with an error function. Figure 5.2(b) shows an instantaneous snap-shot of binary mixing between fluids A and B in a turbulent flow field. The size of the measured volume corresponds to a computational cell in CFD, and it is evident that the average concentration in a computational cell will give a poor description of the concentration during mixing and fast chemical reactions. The theoretical concentrations and mixing rate (scalar dissipation rate) across an interface are presented in Figure 5.2(a) with error functions. At the smallest scales in Figure 5.2(a) only diffusion is important for the mixing process. Turbulence quickly increases the interfacial area between fluid elements and thus reduces the required diffusion distance, so that the required diffusion time will be small. If turbulence is intense, diffusion will limit the rate of mixing, and vice versa if the intensity is too low.

It is of great importance to estimate whether mixing can be a rate-limiting step in the process. The mixing time must then be compared with a timescale for reaction. A quantity that compares the timescale of chemical reactions with the timescale of mixing is the Dahmköhler number (Da):

$$Da = \frac{\text{Typical time required for mixing}}{\text{Typical time required for chemical reactions}}. \tag{5.4}$$

The time for mixing is described in the following equation and the typical time for chemical reactions can be estimated with C_α as the concentration of the limiting reactant:

$$\tau = \frac{C_\alpha}{r} = \frac{C_\alpha}{kC_\alpha C_\beta} = \frac{1}{kC_\beta}.$$

There are three possible outcomes from analysis of the Dahmköhler number:

(1) $Da \ll 1$. Reactions are slow compared with the rate of mixing. In other words, the reactor will have a homogeneous mixture (no segregation) before any substantial reaction can take place. The concentrations in the computational cells are sufficiently well described by the average concentration in the cell. For this scenario standard chemical-reactor models such as the plug-flow reactor or the continuously stirred tank reactor may suffice.

(2) $Da \gg 1$. Reactions are very fast or instantaneous, e.g. acid–base reactions, ion–ion reactions and some gaseous combustion reactions. This problem often benefits from CFD modelling since the common reactor models are inapplicable. Furthermore, since the reactions are instantaneous, it can be shown that studying the evolution of a conserved scalar suffices to get a complete description of the problem (see Section 5.5.2). A conserved scalar is an inert, or passive, tracer, and it is is much easier to study the conserved scalar than reactive species because the (usually nonlinear) chemical reaction rate need not be modelled.

(3) $Da \approx 1$. The timescales for reaction and mixing are of the same order of magnitude. This scenario is by far the most difficult and requires complex modelling for accurate predictions of the mean chemical reaction rates. The case cannot be described by traditional reactor models. Engineers have commonly solved this problem by using Lagrangian micromixing models. However, these models neglect inhomogeneities and reactor geometry.

In modelling of cases (2) and (3) the most successful CFD models calculate first the mixing of a conserved scalar to predict the intensity of segregation of the mixture (the state of mixedness). This information can further be used to obtain closures for the mean chemical reaction rate. When reading Section 5.4 it is important to remember that, even though we are addressing only a conserved scalar, prediction of the mean chemical reaction rate is the overall main objective.

5.4 Mixing of a conserved scalar

To estimate the Dahmköhler number it is necessary to get an idea of the relevant length scales and timescales for mixing and reactions in the system concerned. These scales can be found from simulations of the flow field (the turbulence variables k and ε are usually sufficient) with some model taken from Section 4.2, or from experiments. It is, for example, possible to measure the pressure drop over a chemical reactor and from this estimate the corresponding average rate of energy dissipation. It is also necessary to gain knowledge of the rate expressions for the chemical reactions (like Eq. (5.2)) and the diffusivity of the reacting species. The smallest length scales and timescales that are important for chemical reactions are generally not the same as for the flow (see Chapter 4 about the Kolmogorov scales). This follows since the molecular diffusivity of species in liquids is usually much lower than the kinematic viscosity of the fluid. The smallest relevant length scale for reacting flows will be the average distance a molecule diffuses during the Kolmogorov timescale. This scale is characterized by the Batchelor length scale η_{B}, which is represented as

$$\eta_{\mathrm{B}} = Sc^{-1/2}\eta_{\mathrm{K}}. \tag{5.5}$$

The Schmidt number Sc describes how fast transport of momentum is relative to the transport of molecules ($Sc = \nu/D$). For gases the Schmidt number is approximately unity and the Batchelor scale more or less equals the Kolmogorov scale. For water-like liquids and not-too-large molecules, the Schmidt number is usually close to 1000. The Schmidt number is high for liquids since momentum can be transported by collisions of molecules, whereas molecular diffusion represents the movement of individual molecules.

5.4.1 Mixing timescales

The stages of mixing during which fluid elements are deformed and reduced in size followed by molecular diffusion are referred to in combination as micromixing. The initial step, i.e. deformation and stretching, is referred to as inertial–convective mixing since the fluid elements are merely transported from large eddies to small eddies through convection. The timescale for this process is merely the inverse of the rate of mixing. Hence, modelling of timescales and modelling of mixing rates are equivalent, the timescale used for describing inertial–convective mixing being

$$\tau_{\mathrm{IC}} = \theta\frac{k}{\varepsilon}, \tag{5.6}$$

where θ is a constant commonly set to 0.5. There is no Schmidt-number dependence since viscosity and diffusion are not relevant. For gases ($Sc \approx 1$) this single timescale suffices for a complete description of the mixing process.

At scales just below η_K convective and viscous mixing are of the same order of magnitude, whereas diffusion is still too slow to interfere for $Sc \gg 1$. Mixing on this scale is commonly referred to as engulfment or viscous–convective mixing. The timescale for engulfment is often determined by

$$\tau_{VC} = 17.25\sqrt{\frac{\nu}{\varepsilon}}. \tag{5.7}$$

The final stage of mixing during which all spatial gradients disappear and a homogeneous mixture is obtained occurs near the Batchelor scale. Mixing at this scale is referred to as viscous–diffusive mixing since both viscosity and diffusion are important. The timescale for viscous–diffusive mixing is given as

$$\tau_{VD} = \frac{\tau_{VC}}{0.303 + 17\,050 Sc^{-1}}, \tag{5.8}$$

which is proportional to the viscous–convective timescale, but with a Schmidt-number dependence.

Typical average data for a large stirred-tank reactor with water at room temperature are $k \approx 0.05$ m^2 s^{-2} and $\varepsilon \approx 1$ W kg^{-1}, giving $\tau_{IC} = 25$ ms, $\tau_{VC} = 17$ ms and $\tau_{VD} \approx$ 1 ms. Only fast reactions in which a noticeable amount has reacted within 1 s need to be modelled using mixing models. The mixing time for the reactor, i.e. the time taken to reach the same concentration in the whole reactor, is of the order of minutes, and the ideal stirred-tank reactor may still be a poor model. The local rate of dissipation around the impeller is an order of magnitude higher than the average, and good mixing for fast reactions can be obtained when the reactants are added in the impeller region and the opposite behaviour is obtained close to the reactor surface.

The characterization of timescales for mixing described in this section is only one of several suggestions that have appeared in the literature. However, all mixing models are similar insofar as one must make use of the same parameters that are available from flow-field computations (ν, k, ε and Sc). The important lesson is that you now have some tools that can be used to get a first understanding of the problem at hand, through describing the reactions as slow, fast or instantaneous compared with mixing. You can, for example, use the knowledge from this section to determine the local rate of mixing in a stirred tank, which is important for finding the optimal position for injection of a reactant. Usually it is most efficient to inject where the rate of mixing is fastest, e.g. in the impeller region.

5.4.2 Probability density functions

In modelling of reactive mixing a very important tool is the probability density function (PDF) of a mixture fraction (a conserved scalar, i.e. a non-reacting species). The mixture fraction $\xi(\boldsymbol{x}, t)$ is defined for binary mixtures (mixtures with two inlets) as unity for one inlet stream and zero for the other. The name mixture fraction is thus logical since it

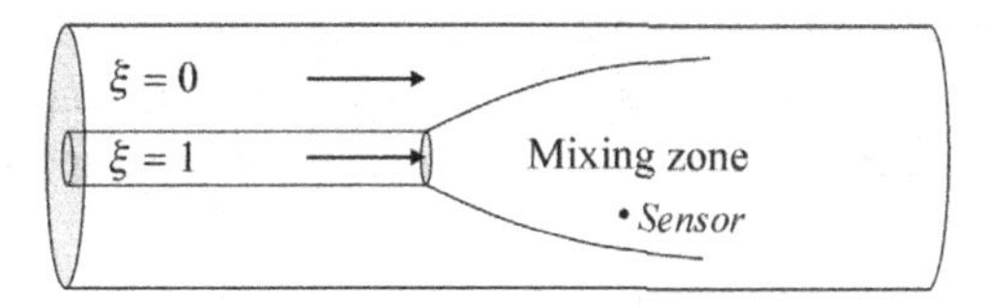

Figure 5.3 Mixing in pipe flow.

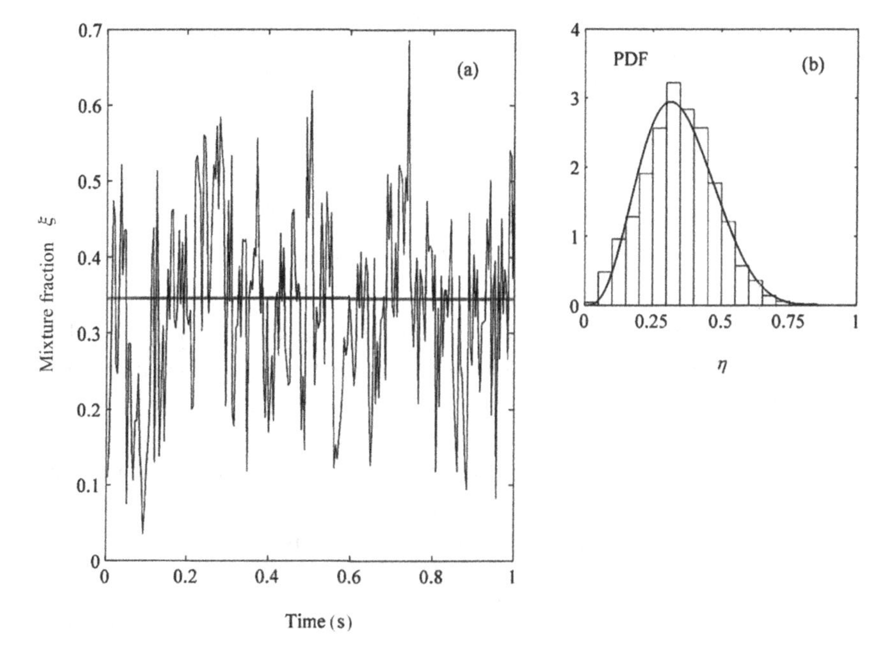

Figure 5.4 Measurements of the mixture fraction at a position in the mixing zone in Figure 5.3. (a) Raw data. (b) Histogram (PDF) of the instantaneous measurements. Here the solid line represents the beta-PDF with mean and variance taken from the experiment. The bars represent the experimental data.

describes how large a fraction of the flow at a certain point has historically come from the injection with a value of unity. Consider an injection in the centre of a turbulent pipe flow (Figure 5.3). This figure shows the instantaneous concentration of the mixture fraction at the injection. If you measure continuously at any (infinitesimally small) point in the pipe, you will obtain a time series of concentrations as shown in Figure 5.4. Figure 5.4(a) shows the raw data with all the characteristics of a chaotic turbulent flow. Like all turbulent measurements, the data will make sense only when some averaging procedure is performed. Figure 5.4(b) shows the histogram, or probability density, of the raw data.

The PDF φ is defined as the probability of measuring a certain concentration η of the tracer:

$$\varphi(\eta)\mathrm{d}\eta \equiv \text{probability of}\,\{\eta \leq \xi \leq \eta + \mathrm{d}\eta\}, \tag{5.9}$$

where η is a sample-space variable for ξ. A sample space is the collection of all possible outcomes of an event. A variable used to describe a single event in sample space is a sample-space variable. In mixture-fraction space η is thus the sample-space variable used to describe the real event ξ.

The sample space of the mixture fraction spans from zero to unity, as shown in Figure 5.4(b), since these two extremes represent pure mixtures. A logical effect of mixing is that the mixture fraction cannot take on values outside this range. For a single bin in the histogram of Figure 5.4(b), $\mathrm{d}\eta$ is chosen to have the discrete value of 0.05, but for a continuous PDF $\mathrm{d}\eta \to 0$. By definition the integral of the PDF must equal unity:

$$\int_0^1 \varphi(\eta)\mathrm{d}\eta = 1. \tag{5.10}$$

For the discrete histogram the integral is defined as the summation over all the bins:

$$\sum_{\text{bins}} \varphi(\eta)\mathrm{d}\eta = 1. \tag{5.11}$$

Another way to think of the PDF is that it describes the fraction of time that the mixture fraction spends in the state η, i.e. the fraction of time a certain concentration or mixture fraction is observed.

The PDF contains all single-point information of the mixture fraction. Given the PDF, all mixture fraction moments (mean, variance, skewness etc.) can be found by integration over mixture-fraction space. The mean $\langle\xi\rangle$ and variance σ^2 (second central moment) of the mixture fraction are defined through the PDF as

$$\langle\xi\rangle = \int_0^1 \eta\varphi(\eta)\mathrm{d}\eta \qquad \text{and} \qquad \sigma^2 = \int_0^1 (\eta - \langle\xi\rangle)^2\varphi(\eta)\mathrm{d}\eta. \tag{5.12}$$

It can be seen from Figure 5.4(b) that for this particular PDF the probability of measuring a mixture fraction close to the mean is relatively large. In other words, the variance is small. What happens during mixing is that the variance will gradually decrease until eventually the mixture is homogeneous, the variance will be zero and the PDF can be described by a single delta-function. In Figure 5.5 the evolution of the PDF for mixing in a homogeneous turbulent flow field with a presumed beta-PDF as a model for φ (the mean mixture fraction will be constant and the variance will decrease exponentially in time) is shown. The beta-PDF uses the mean and variance of the mixture fraction to give a continuous distribution. The beta-PDF φ_B is defined as

$$\varphi_B(\eta; a, b) = \frac{\eta^{a-1}(1-\eta)^{b-1}}{B(a,b)}, \tag{5.13}$$

A semicolon is used to denote that η is a sample-space variable, whereas a and b are fixed parameters. The coefficients are easily calculated from the average ξ and the variance σ^2 as

$$a = \langle\xi\rangle\left[\frac{\langle\xi\rangle(1-\langle\xi\rangle)}{\sigma^2} - 1\right], \qquad b = \frac{1-\langle\xi\rangle}{\langle\xi\rangle}a \tag{5.14}$$

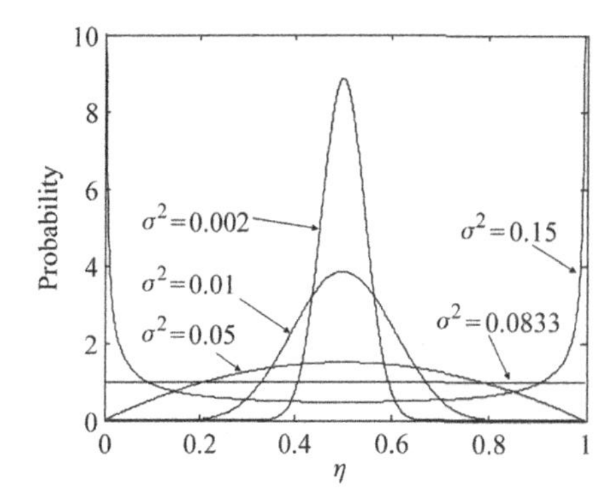

Figure 5.5 Mixing in homogeneous turbulence as described with the beta-PDF. $\langle \xi \rangle = 0.5$.

and $B(a, b)$ is given by the gamma-function Γ that can be found in most mathematical tables:

$$B(a,b) = \int_0^1 s^{a-1}(1-s)^{b-1}\mathrm{d}s = \frac{\Gamma(a)\,\Gamma(b)}{\Gamma(a+b)}. \tag{5.15}$$

It can be seen from Figure 5.5 that the PDF from the experiment nearly coincides with the presumed beta-PDF. This simple observation (that the shape of the beta-PDF nearly resembles the experimentally observed shape of the mixture-fraction PDF) has been used extensively in many closures for reactive flows. There is still no theoretical justification for using the beta-PDF, though, but it is known to give a very good description of mixing in homogeneous flows. For inhomogeneous flows the accuracy is known to be worse, and it is not necessarily a good approximation. Inhomogeneous flows have significant spatial gradients of the calculated properties. In other words, convection and diffusion cannot be neglected in the governing equations. It is generally possible to obtain homogeneous flows only in laboratories or in numerical experiments (DNS).

Nonetheless, the presumed beta-PDF is the most extensively used PDF even for inhomogeneous flows. The advantage with the beta-PDF is that it is a function only of the average ξ and the variance σ^2, and only ξ and σ^2 must be simulated in order to obtain the PDF. Note that the beta-PDF can be accurate only when there are two distinct inlet streams. Consider a case of three separate inlets. One stream has pure tracer ($\xi = 1$), one stream has only water ($\xi = 0$) and the last is a diluted-tracer stream (e.g. a recirculation stream with $\xi = \frac{1}{2}$). The initial PDF will then exhibit three distinct peaks at 0, $\frac{1}{2}$ and 1. The beta-PDF can initially only peak at 0 and 1, and is thus unable to reproduce the experimentally observed PDF. To be able to compute PDFs with multiple peaks, one will have to use more than two moments of the mixture fraction or more than just the one single-mixture fraction. A discussion of this difficult topic is beyond the scope of this book, though.

The presumed beta-PDF has no physical foundation; it is only a convenient description of the mixing, and there are other suggestions for the presumed PDF. The most common are a clipped Gaussian or some ensemble of delta-functions. A presumed PDF that can

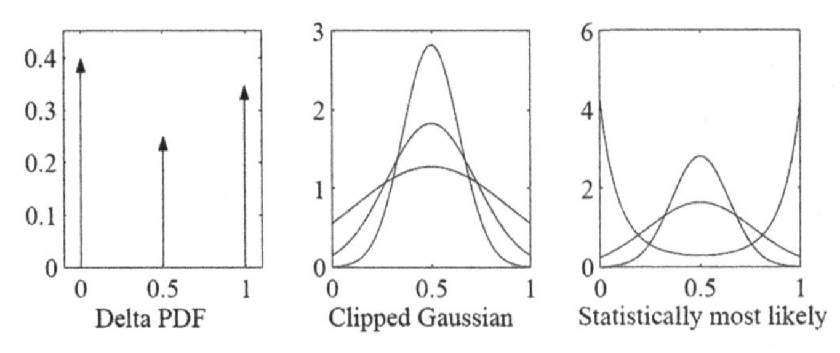

Figure 5.6 Some presumed PDFs. The three realizations of the Gaussian and the statistically most likely PDFs have the same moments ($\langle \xi \rangle = 0.5$, $\langle \xi'^2 \rangle = 0.15, 0.05, 0.02$). The delta PDF has $\langle \xi \rangle = 0.48$ and $\langle \xi'^2 \rangle = 0.19$.

take any number of moments from the mixture fraction is the statistically most likely (SML) PDF [14]. Owing to the possibility of including more moments, the SML PDF can accurately describe even strongly inhomogeneous flows. However, the parameters of the SML PDF have to be found from iterations and thus more computational power is required than for the beta-PDF. Also, the closures for higher moments of the mixture fraction are not nearly as well investigated as those for the mean and the variance. In Figure 5.6 some realizations of the different presumed PDFs are shown.

5.4.3 Modelling of turbulent mixing

The mixture-fraction PDF involves just the first two moments of the mixture fraction, and these moments must be predicted in the mixing model. Transport equations for any moment follow quite simply from manipulations of the transport equation for the instantaneous mixture fraction:

$$\frac{\partial \xi}{\partial t} + U_j \frac{\partial \xi}{\partial x_j} = \frac{\partial}{\partial x_j}\left(D \frac{\partial \xi}{\partial x_j}\right). \tag{5.16}$$

In the simplest case the mixture fraction is merely a normalized concentration and this equation is identical to Eq. (2.1) without the reaction and source terms. We know that Eq. (5.16) can in general not be solved directly, since U_i is unknown, and requires an extremely dense grid resolution for accurate solutions (DNS, see Section 4.2.1). Reynolds decomposition of the velocity and mixture fraction ($U_i = \langle U_i \rangle + u_i$ and $\xi = \langle \xi \rangle + \xi'$) followed by Reynolds averaging of the resulting equation leads to an equation that is more easily solved, but that contains information only on the average mixture fraction:

$$\frac{\partial \langle \xi \rangle}{\partial t} + \langle U_j \rangle \frac{\partial \langle \xi \rangle}{\partial x_j} = -\frac{\partial}{\partial x_j}\left(\langle u_j \xi' \rangle - D \frac{\partial \langle \xi \rangle}{\partial x_j}\right). \tag{5.17}$$

In Eq. (5.17) the first term is accumulation, the second is convection by the mean flow, the third is turbulent transport and the last is molecular diffusion, which usually can be neglected since D is small. The flux $\langle u_j \xi' \rangle$ is unclosed and requires modelling (e.g. Eq. (5.24)). The mean mixture fraction is represented by a thick line in Figure 5.4(a). All detailed information is lost upon this averaging. The mean mixture fraction describes macromixing and gives information about the largest scales of the reactor. It is, however, useless for predicting the level of mixing on the smallest scales (micromixing) which are important for fast or instantaneous chemical reactions.

Micromixing is often described by the mixture-fraction variance which was previously defined through the PDF in Eq. (5.12). The variance can be interpreted as a local average departure from homogeneity, and it is thus a local measure of segregation. With reference to Figure 5.4(a), the variance can be understood as the intensity of the fluctuations around the mean. Strong fluctuations mean a large variance (high degree of segregation). The maximum variance that can be achieved occurs when all measurements are either zero or unity (the existence of no intermediate values means that no mixing has occurred). The value of the variance will then equal $\sigma^2_{\max} = \langle\xi\rangle(1 - \langle\xi\rangle)$, which is often used for normalization. The resulting normalized variance is termed the intensity of segregation I_S:

$$I_S = \frac{\sigma^2}{\langle\xi\rangle(1 - \langle\xi\rangle)}. \tag{5.18}$$

The intensity of segregation always starts with the value of unity under non-premixed initial conditions.

A transport equation for σ^2 can be found by multiplying Eq. (5.16) by 2ξ, Reynolds decomposing U_i and ξ, and then taking the Reynolds average of the whole equation:

$$\frac{\partial \sigma^2}{\partial t} + \langle U_j \rangle \frac{\partial \sigma^2}{\partial x_j} = -\frac{\partial}{\partial x_j}\left(\langle u_j \xi'^2 \rangle - D\frac{\partial \sigma^2}{\partial x_j}\right) - 2\langle u_j \xi' \rangle \frac{\partial \langle\xi\rangle}{\partial x_j} - 2D\left\langle \frac{\partial \xi'}{\partial x_j}\frac{\partial \xi'}{\partial x_j}\right\rangle. \tag{5.19}$$

To derive Eq. (5.19) we have also used Eq. (5.17) and the following identity:

$$\xi \frac{\partial \xi}{\partial x_i} = \frac{1}{2}\frac{\partial \xi^2}{\partial x_i}. \tag{5.20}$$

The left-hand side of Eq. (5.19) is accumulation and convection. On the right-hand side the first term is turbulent transport and molecular diffusion, the second is production due to interaction between a flux and the gradient of the mean mixture fraction, and the last term is twice the mean scalar dissipation rate, which we henceforth denote $\langle N \rangle$ (see Eq. (5.32)). The mean scalar dissipation rate is the scalar analogue of the mean energy-dissipation rate ε and is always positive. Physically, this term describes how fast the mixture-fraction variance is disappearing due to diffusion at the smallest scales.

Note that, even though Eq. (5.19) is an exact equation for the mixture-fraction variance, problems could arise from its solution. The intensity of segregation cannot, by definition, be larger than unity. However, with the solution of Eq. (5.19) there is nothing to ensure that this limit is respected. Problems can arise due to the production term and the closure for the first-order flux $\langle u_j \xi' \rangle$ (see Eq. (5.24)). There is a nice solution to this problem, though. Instead of solving for Eq. (5.19) directly, you can solve for the equivalent second raw moment σ_R^2. The second raw moment is defined as the second moment around zero:

$$\sigma_R^2 = \langle \xi^2 \rangle = \int_0^1 \eta^2 \varphi(\eta) d\eta. \tag{5.21}$$

A transport equation for σ_R^2 can be found by exactly the same procedure as for Eq. (5.19), but without decomposition of ξ. The resulting equation reads

$$\frac{\partial \sigma_R^2}{\partial t} + \langle U_j \rangle \frac{\partial \sigma_R^2}{\partial x_j} = -\frac{\partial}{\partial x_j}\left(\langle u_j \xi^2 \rangle - D\frac{\partial \sigma_R^2}{\partial x_j}\right) - 2D\left\langle \frac{\partial \xi}{\partial x_j}\frac{\partial \xi}{\partial x_j}\right\rangle. \tag{5.22}$$

On implementing Eq. (5.22) instead of Eq. (5.19) you will not get any problems of realizability (unphysical results) since there is no production involved. It also ensures that the intensity of segregation initially starts out as exactly unity. To obtain the central variance from the raw moment, simply apply the following formula:

$$\sigma^2 = \sigma_R^2 - \langle \xi \rangle^2. \tag{5.23}$$

In Eqs. (5.17), (5.19) and (5.22) closure is required for $\langle u_j \xi' \rangle$, $\langle u_j \xi'^2 \rangle$, $\langle u_j \xi^2 \rangle$ and $\langle N \rangle$, so these terms will now be discussed further. The first three terms are fluxes, and are thus required to conserve the mean upon transport (the spatial derivative of the flux is convection, which is conservative). The only way to achieve this is through gradient-diffusion models or by deriving completely new transport equations for the fluxes (that again will contain unknown terms requiring closure). In engineering only the first option is usually employed. The closures for the fluxes thus become

$$\langle u_j \xi' \rangle = -D_T \frac{\partial \langle \xi \rangle}{\partial x_j}, \tag{5.24}$$

$$\langle u_j \xi'^2 \rangle = -D_T \frac{\partial \sigma^2}{\partial x_j} \tag{5.25}$$

$$\langle u_j \xi^2 \rangle = -D_T \frac{\partial \sigma_R^2}{\partial x_j}. \tag{5.26}$$

Here D_T is the turbulent diffusivity, which is commonly calculated as

$$D_T = \frac{\nu_T}{Sc_T}. \tag{5.27}$$

The exact form of D_T will depend on the turbulence model used for ν_T (see Chapter 4). The turbulence Schmidt number Sc_T is known to vary between approximately 0.5 and 1.5, but is most commonly set to 0.7. The near-unity value of Sc_T means that turbulent transport (macromixing) of momentum and species is almost identical for most flows. Note that the gradient closures (Eqs. (5.24) – (5.26)) are closely related to the Boussinesq hypothesis used for modelling of the Reynolds stresses discussed in Section 4.2.4.

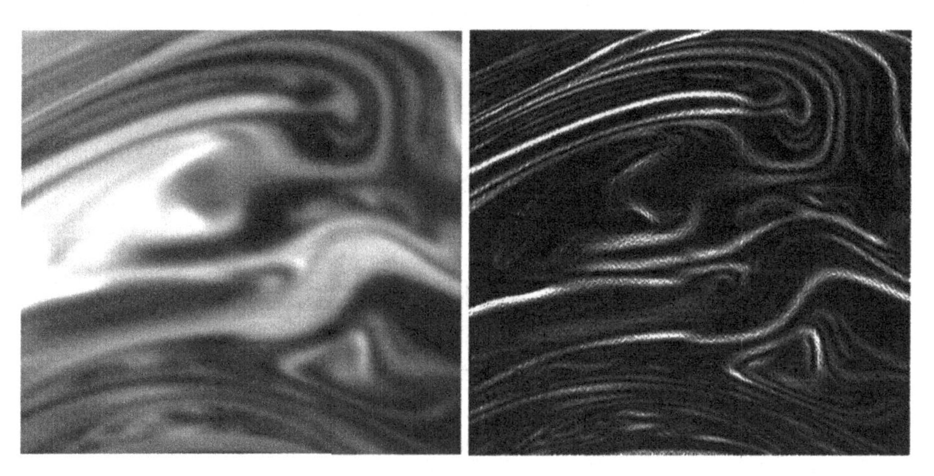

Figure 5.7 Concentration fluctuations (left) and scalar dissipation (right).

The closed models for the average mixture fraction, the variance and the raw variance are then obtained as

$$\frac{\partial \langle \xi \rangle}{\partial t} + \langle U_j \rangle \frac{\partial \langle \xi \rangle}{\partial x_j} = \frac{\partial}{\partial x_j} \left((D + D_{\mathrm{T}}) \frac{\partial \langle \xi \rangle}{\partial x_j} \right), \tag{5.28}$$

$$\frac{\partial \sigma^2}{\partial t} + \langle U_j \rangle \frac{\partial \sigma^2}{\partial x_j} = \frac{\partial}{\partial x_j} \left((D + D_{\mathrm{T}}) \frac{\partial \sigma^2}{\partial x_j} \right) + 2D_{\mathrm{T}} \frac{\partial \langle \xi \rangle}{\partial x_j} \frac{\partial \langle \xi \rangle}{\partial x_j} - 2D \left\langle \frac{\partial \xi'}{\partial x_j} \frac{\partial \xi'}{\partial x_j} \right\rangle, \tag{5.29}$$

$$\frac{\partial \sigma_{\mathrm{R}}^2}{\partial t} + \langle U_j \rangle \frac{\partial \sigma_{\mathrm{R}}^2}{\partial x_j} = \frac{\partial}{\partial x_j} \left((D + D_{\mathrm{T}}) \frac{\partial \sigma_{\mathrm{R}}^2}{\partial x_j} \right) - 2D \left\langle \frac{\partial \xi}{\partial x_j} \frac{\partial \xi}{\partial x_j} \right\rangle. \tag{5.30}$$

The most important term to close in turbulent mixing is the mean scalar dissipation rate. In Figure 5.2(a) we have already shown a realization of the instantaneous scalar dissipation rate. When the instantaneous scalar dissipation is sampled at the same point over a sufficiently long period of time, we obtain the mean. Note that the mean scalar dissipation rate is the only term, besides transport, that requires closure for turbulent-mixing problems. The scalar dissipation rate describes how fast the final small-scale mixing occurs, or how fast we are obtaining a homogeneous mixture. In Figure 5.2 we have shown the interface between two fluids undergoing mixing. The scalar dissipation rate is defined as the molecular diffusivity times the square of the gradient of the scalar:

$$N = D \frac{\partial \xi}{\partial x_j} \frac{\partial \xi}{\partial x_j}. \tag{5.31}$$

Since the magnitude of the scalar gradient is largest at the centre of the interface, this is where mixing will be fastest. Figure 5.7 shows that the interfaces can be very narrow in high-Schmidt-number liquids, which is why mixing is often referred to as an intermittent phenomenon. By this we mean that a very small part of the total volume of a reactor actually contributes to the final micromixing. Since mixing is fastest at the centre of an

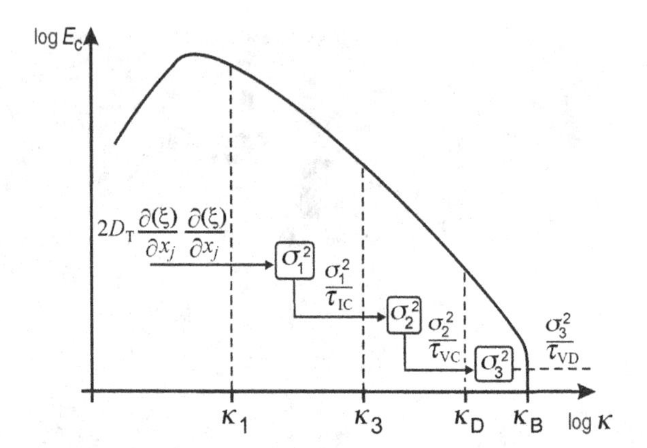

Figure 5.8 The variance cascade in the TMM model.

interface, this is also where the rate of fast chemical reactions (mixing controlled) will peak.

The scalar dissipation rate is more difficult to close for liquids than gases since there is a large separation of scales and any one of inertial–convective, viscous–convective and viscous–diffusive mixing can be rate limiting (see Section 5.4.1). The simplest closures for $\langle N\rangle$ employ a mixing-frequency closure

$$\langle N\rangle = \frac{\sigma^2}{2\tau}, \tag{5.32}$$

where τ is usually calculated as the inertial–convective timescales defined in Eq. (5.6). (Any of the other timescales can still be employed, though.) If this method for predicting $\langle N\rangle$ is too crude, it is possible to obtain higher accuracy through deriving an exact transport equation for $\langle N\rangle$. In practice this is rarely justified, though.

To further complicate modelling of $\langle N\rangle$, the scalar is injected on an inertial scale that usually differs from the largest scales of the flow. Consider again injection at the centre of a turbulent pipe flow (see Figure 5.3). The injected scalar enters the pipe with eddies whose sizes are determined by the radius of the injection pipe. This pipe is, however, much smaller than the radius of the outer pipe, which determines the largest eddies in the main flow. Consequently, close to the injection there will be a transitional region where variance is produced and the largest scale of mixing approaches that of the main flow. To be able to describe this transitional region, a dynamical multi-scale model must be employed. Such a model has been described in detail by Fox [15]. A more intuitive model denoted the turbulent mixer model (TMM) has been described by Baldyga [14] and is presented here. The TMM assumes that the local value of σ^2 can be divided into three parts according to the scales of segregation, namely inertial–convective (σ_1^2), viscous–convective (σ_2^2) and viscous–diffusive (σ_3^2):

$$\sigma^2 = \sigma_1^2 + \sigma_2^2 + \sigma_3^2. \tag{5.33}$$

Further, it is assumed that variance is produced on the macrolevel and dissipated to the smaller scales in a cascade as shown in Figure 5.8. Hence dissipation of σ_1^2 leads to

Table 5.1 Constants in the turbulent mixer model

α	P_α	D_α
1	$2D_T \dfrac{\partial \langle \xi \rangle}{\partial x_j} \dfrac{\partial \langle \xi \rangle}{\partial x_j}$	$\dfrac{\sigma_1^2}{\tau_{IC}}$
2	$\dfrac{\sigma_1^2}{\tau_{IC}}$	$\dfrac{\sigma_2^2}{\tau_{VC}}$
3	$\dfrac{\sigma_2^2}{\tau_{VC}}$	$\dfrac{\sigma_3^2}{\tau_{VD}}$

production of σ_2^2 and dissipation of σ_2^2 leads to production of σ_3^2. Dissipation of σ_3^2, on the other hand, leads to complete mixing. The transport equations for all variances are given as

$$\frac{\partial \sigma_\alpha^2}{\partial t} + \langle U_j \rangle \frac{\partial \sigma_\alpha^2}{\partial x_j} = \frac{\partial}{\partial x_j}\left[(D + D_T)\frac{\partial \sigma_\alpha^2}{\partial x_j}\right] + P_\alpha - D_\alpha \qquad \text{for} \quad \alpha = 1, 2, 3 \tag{5.34}$$

and the production P_α and dissipation D_α terms are given in Table 5.1. The molecular-diffusion term in Eq. (5.34) is usually negligible since in general $D_T \gg D$.

The first three terms of Eq. (5.34) account merely for accumulation and transport. The difficult term here is the sink term for the variance, i.e. the scalar dissipation. In the TMM model it is set as the variance divided by the relevant time for each scale.

By summation of the three transport equations and subtraction of Eq. (5.19) we find that the mean scalar dissipation rate is given indirectly as

$$\langle N \rangle = \frac{\sigma_3^2}{2\tau_{VD}}. \tag{5.35}$$

The variance describes the concentration fluctuations in time and at the inlets. With constant concentration all central variances have initial values of zero.

Direct implementations of the TMM will usually lead to realizability problems due to the production term P_1. As discussed before, the solution to this problem is to use raw, instead of central, moments for the inertial–convective variance σ_1^2. For the small-scale variances (σ_2^2 and σ_3^2) there will be no problems of realizability, since the production here does not include spatial gradients. Hence they are not subject to discretization errors. To be able to use the TMM for raw moments, simply omit P_1 and set the appropriate initial conditions (σ_{R1}^2, $= 1$, where $\langle \xi \rangle = 1$ according to Eq. (5.21), but still with $\sigma_2^2 = \sigma_3^2 = 0$). The dissipation term D_1 must still contain the central variance, now expressed as $\sigma_1^2 = \sigma_{R1}^2 - \langle \xi \rangle^2$. The true raw moment can be recovered as

$$\sigma_R^2 = \sigma_{R1}^2 + \sigma_2^2 + \sigma_3^2 \tag{5.36}$$

and the central moment through Eq. (5.23).

In this section we have discussed only pure mixing. By solving the equations above, we obtain the local mean mixture fraction and variance. The intention has been to obtain a tool for predicting the turbulence-chemistry effect in order to be able to predict the average rate of chemical reactions in each computational cell. In the next section we will discuss how the modelling of a conserved scalar is used in reference to this.

5.5 Modelling of chemical reactions

The modelling hierarchy for reactive flows closely resembles that of pure flows (see Chapter 4). In general you have to give up information in return for computational efficiency. In this regard it is important to understand what information you are giving up and how this can be modelled through simplified relations. As for turbulence modelling the word 'simplified' is actually very misleading. The problem is not easier than direct numerical simulation (DNS) of the governing equations. However, DNS is possible only for simple geometries and low Reynolds numbers. Hence, DNS is not an option for chemical reactors and will most likely remain just a research tool in the foreseeable future.

The instantaneous equation for reacting species that is solved in DNS consists of standard accumulation, convection, diffusion and reaction terms:

$$\frac{\partial C_\alpha}{\partial t} + U_j \frac{\partial C_\alpha}{\partial x_j} = \frac{\partial}{\partial x_j}\left(D \frac{\partial C_\alpha}{\partial x_j}\right) + S_\alpha(\mathbf{C}). \tag{5.37}$$

Here **C** denotes a vector of all the reacting species in the flow, meaning that the source term for species α, S_α, could depend on any of the other existing species. Since Eq. (5.37) can be solved only for simple geometries with extremely dense meshes (DNS), we must introduce some sort of averaging. Here we consider only Reynolds averaging.

Consider again the reaction described in Eq. (5.1). For a reaction rate $r = kC_\mathrm{A}C_\mathrm{B}$ the characteristic chemical-reaction timescale can now be given as either τ_R = minimum of $1/(k_1 C_\mathrm{A})$ and $1/(k_1 C_\mathrm{B})$ or $\tau_\mathrm{R} = 1/(k_1 C_\mathrm{A} + k_1 C_\mathrm{B})$, where the concentrations are defined at the inlets. (Note that, since this is only an approximate timescale, the definitions are not strict.) This timescale should now be used to predict the Dahmköhler number to see which scenario of Section 5.3 is relevant.

5.5.1 *Da* $\ll 1$

If the Dahmköhler number is small there is no problem, since there is sufficient time for local mixing and the covariance term in Eq. (5.3) will be zero. The reaction rate can be expressed merely through the mean concentrations that are already known. The reaction described in Eqs. (5.1)–(5.3) can now easily be closed by neglecting the covariance in Eq. (5.3), and we obtain

$$S_\mathrm{A} = -k_1 C_\mathrm{A} C_\mathrm{B}. \tag{5.38}$$

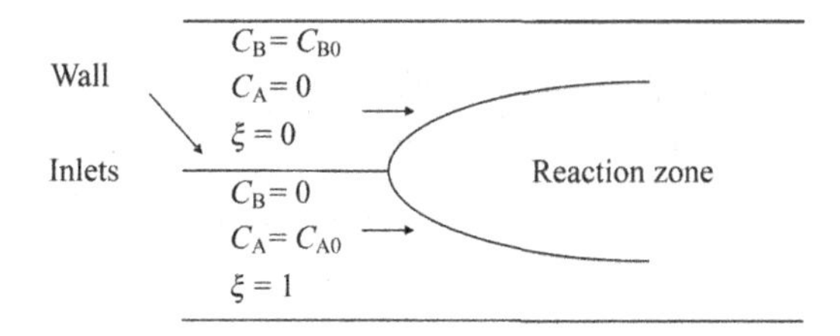

Figure 5.9 A reactive-mixing layer. The reacting species are initially separated by a wall.

The concentrations for A and B are then obtained from the following equations that are easily solved by the CFD software:

$$\frac{\partial\langle C_A\rangle}{\partial t} + U_j\frac{\partial\langle C_A\rangle}{\partial x_j} = \frac{\partial}{\partial x_j}\left[(D + D_T)\frac{\partial\langle C_A\rangle}{\partial x_j}\right] + \gamma_A k\langle C_A\rangle\langle C_B\rangle \tag{5.39}$$

and

$$\frac{\partial\langle C_B\rangle}{\partial t} + U_j\frac{\partial\langle C_B\rangle}{\partial x_j} = \frac{\partial}{\partial x_j}\left[(D + D_T)\frac{\partial\langle C_B\rangle}{\partial x_j}\right] + \gamma_B k\langle C_A\rangle\langle C_B\rangle. \tag{5.40}$$

5.5.2 $Da \gg 1$

The reaction rate defined in Eq. (5.3) is

$$\langle S_A\rangle = \langle S_B\rangle = -\langle k_1 C_A C_B\rangle = -k_1(\langle C_A\rangle\langle C_B\rangle + \langle C_A' C_B'\rangle). \tag{5.41}$$

For instantaneous reactions, k_1 will have a very large value and the only way to obtain finite reaction rates is if we have

$$-\langle C_A' C_B'\rangle \approx \langle C_A\rangle\langle C_B\rangle. \tag{5.42}$$

Hence in this case the average reaction rate will be approximately zero, which is possible only if the diffusion distance shown in Fig. 5.2(a) is infinitely short. Equation (5.42) is not a closure, though. The closure for instantaneous reactions assumes that A and B cannot coexist in fluid elements. It will be shown below that it is then sufficient to calculate the mixture-fraction PDF (see Section 5.4.2) to completely describe the reacting system. Physically the rate of instantaneous reactions will peak at the centre of a mixing layer (see Figure 5.2), since this is where mixing is most intense.

Consider the instantaneous reaction between species A and B, for any combination of A and B and for any rate expression S:

$$\gamma_A \mathrm{A} + \gamma_B \mathrm{B} \xrightarrow{k_1} \text{products}. \tag{5.43}$$

Here γ_A and γ_B determine the stoichiometry and the initial conditions are given in Figure 5.9.

The instantaneous transport equations (not Reynolds averaged) for A and B read

$$\frac{\partial C_A}{\partial t} + U_j\frac{\partial C_A}{\partial x_j} = \frac{\partial}{\partial x_j}\left(D\frac{\partial C_A}{\partial x_j}\right) + \gamma_A S \tag{5.44}$$

and

$$\frac{\partial C_B}{\partial t} + U_j \frac{\partial C_B}{\partial x_j} = \frac{\partial}{\partial x_j}\left(D \frac{\partial C_B}{\partial x_j}\right) + \gamma_B S, \tag{5.45}$$

where we have assumed that A and B have the same molecular diffusivity D. Multiplying Eq. (5.44) by γ_B and Eq. (5.45) by γ_A and then subtracting Eq. (5.45) from Eq. (5.44) leads to

$$\frac{\partial(\gamma_B C_A - \gamma_A C_B)}{\partial t} + U_j \frac{\partial(\gamma_B C_A - \gamma_A C_B)}{\partial x_j} = \frac{\partial}{\partial x_j}\left(D \frac{\partial(\gamma_B C_A - \gamma_A C_B)}{\partial x_j}\right). \tag{5.46}$$

By inspection it can be seen that the constructed variable $(\gamma_B C_A - \gamma_A C_B)$ is satisfied by the same transport equation as the mixture fraction (see Eq. (5.16)) since the source term has disappeared. To be completely identical, the constructed variable must also have the same initial values as the mixture fraction. Normalization leads to a coupling between reactive scalars and the mixture fraction:

$$\xi = \frac{\gamma_B C_A - \gamma_A C_B + \gamma_A C_{B0}}{\gamma_B C_{A0} + \gamma_A C_{B0}}. \tag{5.47}$$

On substituting for $(\gamma_B C_A - \gamma_A C_B)$ from Eq. (5.46) we obtain Eq. (5.16). The inlet containing $C_A = C_{A0}$ and $C_B = 0$ gives $\xi = 1$, and for the inlet containing $C_B = C_{B0}$ and $C_A = 0$ we obtain $\xi = 0$. It is not possible to solve Eq. (5.16) for turbulent flows, but an estimation of the instantaneous mixture fractions can be done using the beta-PDF. By taking the Reynolds average in Eq. (5.28) and solving for the mixture-fraction average $\langle \xi \rangle$ and the variance in Eq. (5.29) or Eq. (5.34), we can reconstruct the beta-PDF using Eqs. (5.13)–(5.15).

For an instantaneous irreversible reaction this relation is especially favourable, since we know that A and B cannot coexist in a fluid element. So, if $C_A > 0$, we know that $C_B = 0$ and vice versa. We also know that there must be a point in mixture-fraction space where $C_A = C_B = 0$. This point is termed the stoichiometric mixture fraction ξ_s and is given by Eq. (5.47) as

$$\xi_s = \frac{\gamma_A C_{B0}}{\gamma_B C_{A0} + \gamma_A C_{B0}}. \tag{5.48}$$

On dividing Eq. (5.47) by Eq. (5.48) we can construct linear correlations between the reactive concentrations and the mixture fraction for all mixture fractions:

$$\gamma_B C_A - \gamma_A C_B = \gamma_A C_{B0}\left(\frac{\xi}{\xi_s} - 1\right). \tag{5.49}$$

We do not know the exact value of ξ, but we can estimate a possible distribution of ξ. In other words, the reactive concentrations are given *conditional* on the value of the mixture fraction. Since these concentrations are constructed in mixture-fraction space, we use the sample-space variable η for ξ.

For $\eta \leq \xi_s$, $C_A(\eta) = 0$ and

$$C_B(\eta) = C_{B0}\left(1 - \frac{\eta}{\xi_s}\right). \tag{5.50}$$

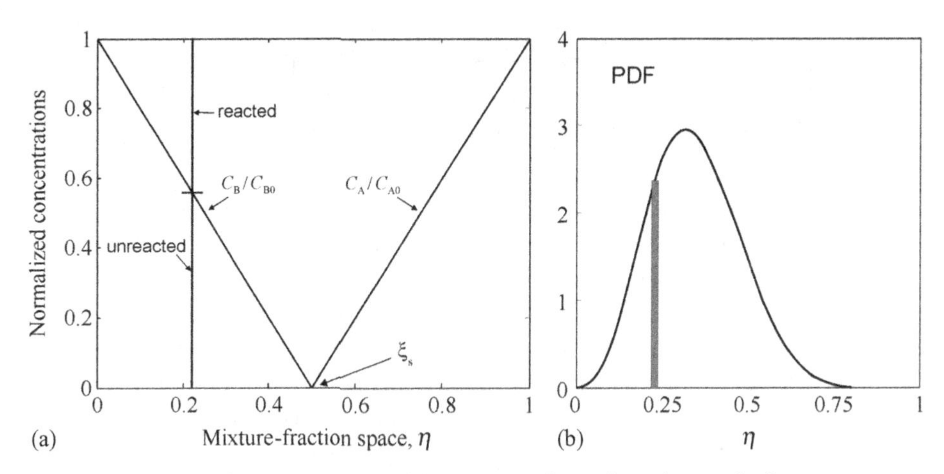

Figure 5.10 A schematic representation of the concentrations of reactive species in mixture-fraction space. The reaction between A and B is instantaneous.

For $\eta \geq \xi_s$, $C_B(\eta) = 0$ and

$$C_A(\eta) = C_{B0}\frac{\gamma_A}{\gamma_B}\left(\frac{\eta}{\xi_s} - 1\right). \tag{5.51}$$

The concentrations of the reactive species are shown schematically in Figure 5.10(a). The total amount of A and B can then be calculated if we know the mixture-fraction distribution $\varphi(\eta)$ in the computational cell shown in Figure 5.10(b). The average concentration in the cell is the integration of the concentration at a given mixture fraction η times the frequency of the appearance of that mixture fraction, i.e. the PDF $\varphi(\eta)$. The mean concentrations can easily be calculated from Eqs. (5.50) and (5.51) weighted by the probability of finding that mixture fraction estimated from the presumed PDF of the mixture fraction:

$$\langle C_A \rangle = \int_0^1 C_A(\eta)\varphi(\eta)\mathrm{d}\eta = C_{B0}\frac{\gamma_A}{\gamma_B}\int_{\xi_s}^1 \left[\frac{\eta}{\xi_s} - 1\right]\varphi(\eta)\mathrm{d}\eta \tag{5.52}$$

and

$$\langle C_B \rangle = \int_0^1 C_B(\eta)\varphi(\eta)\mathrm{d}\eta = C_{B0}\int_0^{\xi_s} \left[1 - \frac{\eta}{\xi_s}\right]\varphi(\eta)\mathrm{d}\eta. \tag{5.53}$$

The only term that needs to be modelled in Eqs. (5.52) and (5.53) is the mixture-fraction PDF $\varphi(\eta)$, which can be closed with, for example, the beta-PDF (see Section 5.4.2). Hence, for instantaneous reactions there is no need to calculate a mean reaction rate explicitly and there is no need for transport equations of the reactive species. For instantaneous reactions 'mixed is reacted' is valid, and only the fraction not mixed need be calculated. The progress of the reaction is simply obtained by calculating the average concentrations $\langle C_A \rangle$ and $\langle C_B \rangle$ along the reactor. It is important to realize that

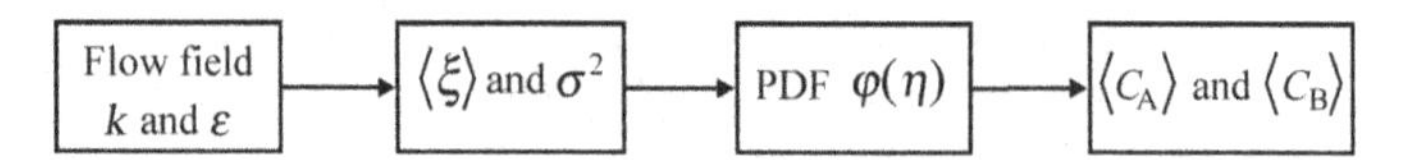

Figure 5.11 The implementation strategy for instantaneous reactions.

the expressions in Eqs. (5.50) and (5.51) follows from the instantaneous-reaction rate assumption that A and B cannot coexist in fluid elements.

Figure 5.11 shows the computational procedure to simulate the average concentration of A and B. First, the flow field is simulated using standard turbulence models. The PDF is obtained from simulation of the average mixture fraction and variance, and finally the average concentrations are obtained from integration of Eqs. (5.52) and (5.53).

In many fast reactions thermodynamics limits the conversion, but these reactions can also be simulated using the PDF method. The only restrictions are that there is no kinetic limitation and that the concentrations of reactants and products can be calculated as functions of the mean mixture fraction $\langle\xi\rangle$ and the variance σ^2. For an isothermal or adiabatic reaction it is possible to calculate the chemical composition and temperature corresponding to the minimum Gibbs energy for all mean mixture fractions and variances. These calculations will require large computational power if they are done in each iteration, but, since the calculations depend only on the mean mixture fraction, the variance and the inlet conditions, it is possible to do the calculations in advance and store them in a look-up table.

The look-up table can be constructed by assuming an equilibrium reaction written as

$$\gamma_A \mathrm{A} + \gamma_B \mathrm{B} \rightleftarrows \gamma_C \mathrm{C} + \gamma_D \mathrm{D} \tag{5.54}$$

with the reaction rate

$$S = k^+ C_A C_B - k^- C_C C_D, \tag{5.55}$$

where k^+ and k^- are the forward and backward reaction rate constants, respectively. On inserting this reaction rate into Eqs. (5.37) and (5.38) a mixture fraction can be formed by subtracting one of these two equations from the other:

$$\xi = \frac{\gamma_B C_A - \gamma_A C_B + \gamma_A C_B^0}{\gamma_B C_A^0 + \gamma_A C_B^0}. \tag{5.56}$$

We can also write transport equations for the products

$$\frac{\partial C_C}{\partial t} + U_j \frac{\partial C_C}{\partial x_j} = \frac{\partial}{\partial x_j}\left(D\frac{\partial C_C}{\partial x_j}\right) + \gamma_C S \tag{5.57}$$

and

$$\frac{\partial C_D}{\partial t} + U_j \frac{\partial C_D}{\partial x_j} = \frac{\partial}{\partial x_j}\left(D\frac{\partial C_D}{\partial x_j}\right) + \gamma_D S. \tag{5.58}$$

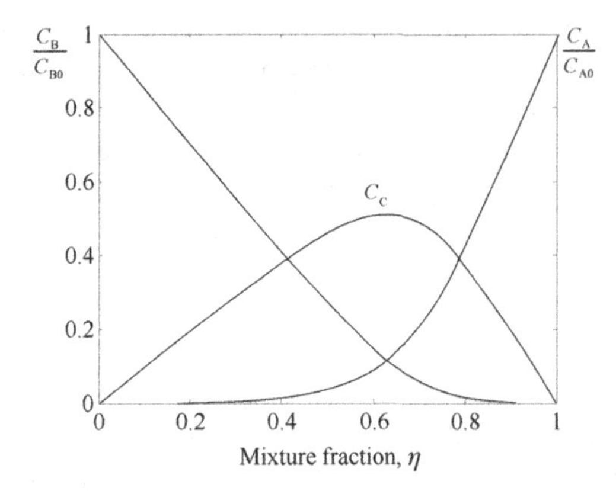

Figure 5.12 Concentrations as functions of the mixture fraction for an equilibrium reaction assuming that $C_{B0} = 2C_{A0}$.

After multiplying by stoichiometric coefficients and adding Eqs. (5.37) and (5.57), we obtain

$$\frac{\partial(\gamma_C C_A + \gamma_A C_C)}{\partial t} + U_j \frac{\partial(\gamma_C C_A + \gamma_A C_C)}{\partial x_j} = \frac{\partial}{\partial x_j}\left(D \frac{\partial(\gamma_C C_A + \gamma_A C_C)}{\partial x_j}\right). \quad (5.59)$$

On defining the mixture fraction

$$\xi = \frac{\gamma_C C_A + \gamma_A C_C}{\gamma_C C_A^0}, \quad (5.60)$$

we obtain Eq. (5.16). Since the mixture fractions formulated with Eq. (5.56) and Eq. (5.60) are identical and have the same boundary conditions, $\xi = 1$ for $C_A = C_{A0}$ and $\xi = 0$ for $C_B = C_{B0}$, they have the same solution and hence

$$\xi = \frac{\gamma_C C_A + \gamma_A C_C}{\gamma_C C_A^0} = \frac{\gamma_B C_A - \gamma_A C_B + \gamma_A C_B^0}{\gamma_B C_A^0 + \gamma_A C_B^0} \quad (5.61)$$

everywhere. For an equilibrium reaction the equilibrium balance

$$\frac{C_C^{\gamma_C} C_D^{\gamma_D}}{C_A^{\gamma_A} C_B^{\gamma_B}} = K(T) \quad (5.62)$$

must be fulfilled and the equilibrium constant is obtained from the thermodynamics:

$$K(T) = e^{\Delta S/R} e^{-\Delta H/(RT)} = e^{-\Delta G/(RT)}, \quad (5.63)$$

where ΔS is the entropy, ΔH is the enthalpy and ΔG is the Gibbs free energy for the reaction. For an adiabatic reaction, the temperature will depend on the inlet temperature and the heat of reaction:

$$T = T_1 + (T_2 - T_1)\xi + \frac{(-\Delta H)C_C}{\gamma_C \rho C_p}, \quad (5.64)$$

where T_1 is the temperature at $\xi = 0$ and T_2 is that at $\xi = 1$. Here it is assumed that the heat capacity is equal for each compound and also constant over the temperature range. The four equations (5.61)–(5.64) contain four unknowns. Hence all the concentrations and the temperature can be calculated as a function of the mixture fraction. Figure 5.12

shows how C_A, C_B and C_C vary with the mixture fraction for a typical equilibrium reaction. For non-adiabatic reactions, the heat loss must be integrated along the reaction path.

The average concentration and temperature are obtained by integrating the instantaneous variations shown in Figure 5.12 using the PDF $\varphi(\eta)$:

$$\langle C_A \rangle = \int_0^1 C_A(\eta)\varphi(\eta)d\eta. \tag{5.65}$$

The PDF is obtained from the simulated mean mixture fraction $\langle \xi \rangle$ and variance σ^2 shown in Figure 5.10(b) using the beta-PDF. The data in Figure 5.12 are obtained from the inlet conditions and thermodynamics, and the average concentrations are functions only of the mean mixture fraction $\langle \xi \rangle$ and variance σ^2. Consequently, it is possible to pre-calculate average concentrations and temperature in a 2D look-up table. This method also allow predictions of radical species' concentrations and dissociation effects at high temperature without knowing the reaction rate, since an equilibrium condition is assumed. The cost of doing these predictions comes directly from the cost of generating a more advanced look-up table. An important limitation with excluding the kinetics is that we cannot predict ignition or extinction of reactions in combustion systems. If these phenomena are of interest, there is no way around the problem and the kinetics must be included in the simulations.

The assumption that there are no kinetic limitations and that the reaction is determined only by thermodynamics is very important. According to the equilibrium model methane can burn at room temperature. Another common problem is when the equilibrium changes very much with temperature, e.g. NO is formed from oxidation of nitrogen above 1500 K and reaches high concentrations in internal combustion engines above 2000 K. The equilibrium model will predict a reversible reaction back to nitrogen and oxygen during the fast cooling due to expansion in the cylinder. In reality this will not happen, since the reaction rate for decomposition of NO is very slow below 2000 K. It is recommended that nitrogen oxides should not be included in the equilibrium calculations if the temperature is below 2000 K. However, formation of nitrogen oxides can often be added as a homogeneous reaction since nitrogen and oxygen are already premixed in the air.

Figure 5.13 shows the different steps in modelling the oxidation of methane in air. First the look-up table is calculated. In Figures 5.13(c) and (f) the mole fraction of methane and temperature are visualized, but a look-up table will be calculated for all compounds that you selected, e.g. CO_2, CO, H_2O, H, CH etc. Note that the variable ξ ranges from 0 to 1 and σ^2 ranges from 0 to 0.25, which is the theoretical maximal variance of unmixed reactants. In the second step the flow, mean mixture fraction and variance are simulated. The temperature and composition can then be found in the look-up tables to obtain the right properties of the fluid. The mean mixture fraction, variance, methane mole fraction and temperature are shown in Figures 5.13(c)–(f). Note that the conserved scalar disappears due to dilution whereas the reactant disappears due to combustion.

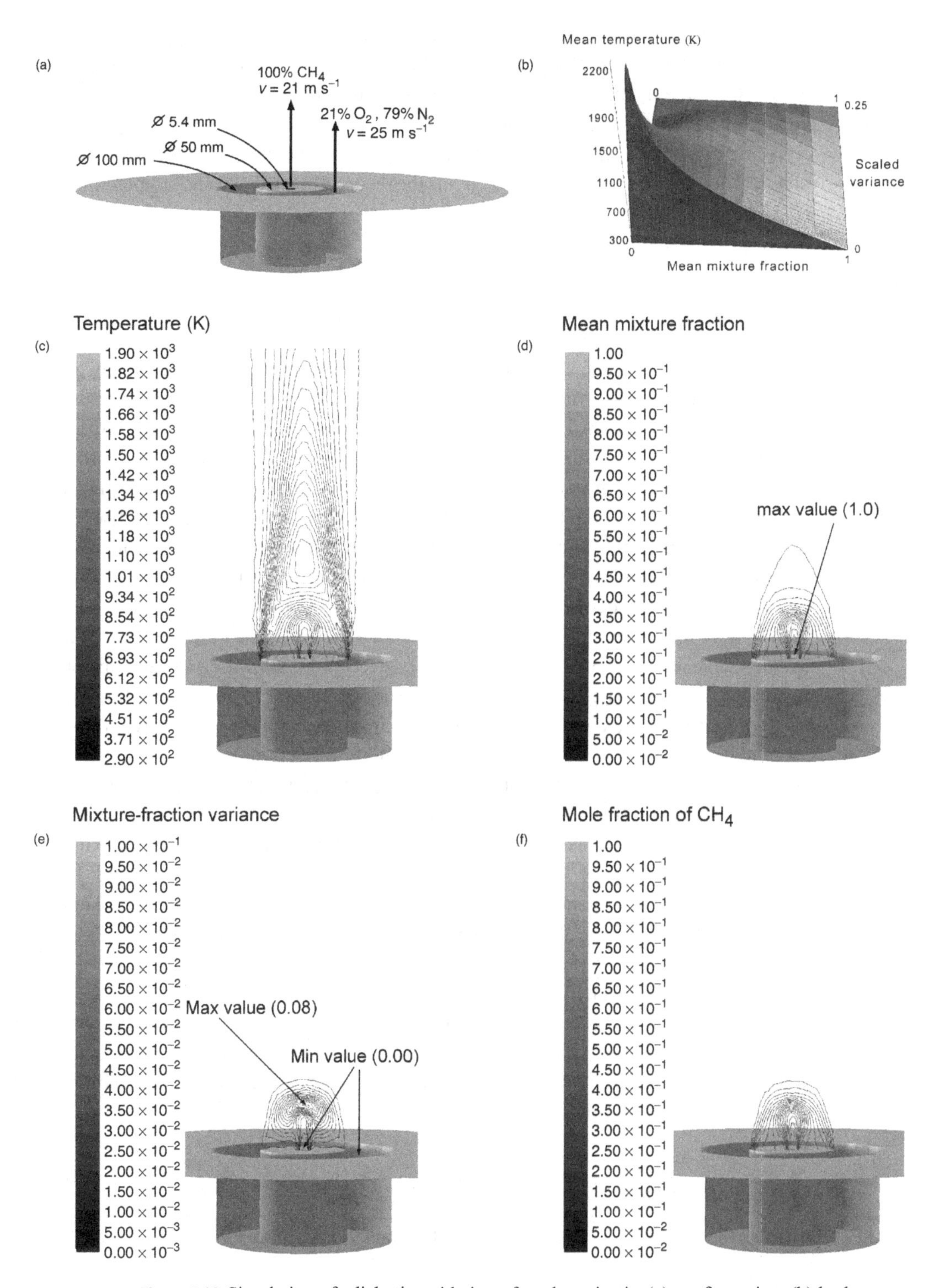

Figure 5.13 Simulation of adiabatic oxidation of methane in air: (a) configuration, (b) look-up table, (c) temperature, (d) mean mixture fraction, (e) variance and (f) mole fraction of CH_4.

It is also possible to use such methods for non-adiabatic reactions. The compositions and temperatures in the look-up tables will then be functions of energy loss or gain also. During the iterations the energy loss or gain is estimated, which allows convection and radiation from the flame to be included in the simulations.

As mentioned before, the PDF methods have been developed only for two distinct inlet streams. For more complicated systems the theories break down and simpler models must be employed.

5.5.3 $Da \approx 1$

For intermediate Dahmköhler numbers A and B can coexist, and this case is by far the most difficult to close. The very simplest models which are able to give reasonably good a-priori predictions use a reaction-progress variable in conjunction with the presumed PDF of a mixture fraction. The reaction-progress variable models are similar in form to the solution presented for instantaneous reactions in Section 5.5.2. The only significant difference is that a transport equation needs to be solved for the reaction-progress variable, which in turn requires the explicit calculation of the average reaction rate. Owing to the mathematical complexity of these models, we will in this section present only one such solution for a simple reaction scheme. For a complete review of today's state of the art, the reader is referred to [15].

To be able to predict mean concentrations with a presumed PDF method, the instantaneous concentrations of the reactive species need to be known or closed subject to the mixture fraction. The instantaneous concentration subject to the mixture fraction is usually referred to as a conditional average. For instantaneous reactions the conditional concentrations have been shown to be linear in mixture-fraction space and hence closed (see Eqs. (5.50) and (5.51)). For intermediate Da, the assumption leading to Eqs. (5.50) and (5.51) does not hold and other relationships must be found. Baldyga and Bourne [14] described a model that uses piecewise linear interpolation between the extreme limits of instantaneous ($k_1 = \infty$) and slow ($k_1 = 0$) chemistry to obtain closure. Here we present this model, again for the simplest case of one fast, but irreversible, reaction as illustrated in Eq. (5.43). It is possible, though, to extend the model to more complex reaction systems.

As has already been stated, the instantaneous rate assumption fails for intermediate Da, so it is necessary to find alternative closures for Eqs. (5.50) and (5.51). The interpolation model exploits the fact that the conditional concentrations have to fall within the extremes of instantaneous and slow, for which the solutions are known. Hence the conditional concentration of A is limited according to

$$\frac{C_{\mathrm{A}}^{0}}{C_{\mathrm{A0}}} \geq \frac{C_{\mathrm{A}}}{C_{\mathrm{A0}}} \geq \frac{C_{\mathrm{A}}^{\infty}}{C_{\mathrm{A0}}} \Rightarrow \eta \geq \frac{C_{\mathrm{A}}}{C_{\mathrm{A0}}} \geq \left(\eta \xi_{\mathrm{s}}^{-1} - 1\right) \frac{\gamma_{\mathrm{A}} C_{\mathrm{B0}}}{\gamma_{\mathrm{B}} C_{\mathrm{A0}}}, \tag{5.66}$$

whereas the conditional concentration of B is limited according to

$$\frac{C_{\mathrm{B}}^{0}}{C_{\mathrm{B0}}} \geq \frac{C_{\mathrm{B}}}{C_{\mathrm{B0}}} \geq \frac{C_{\mathrm{B}}^{\infty}}{C_{\mathrm{B0}}} \Rightarrow 1 - \eta \geq \frac{C_{\mathrm{B}}}{C_{\mathrm{B0}}} \geq 1 - \eta \xi_{\mathrm{s}}^{-1}. \tag{5.67}$$

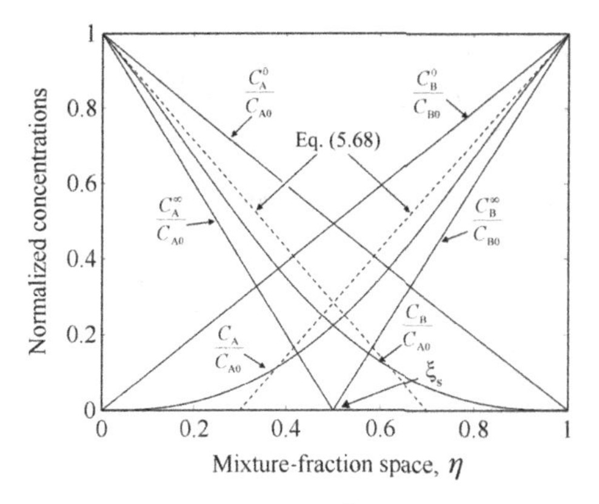

Figure 5.14 Concentrations of reactive species in mixture-fraction space. The reaction between A and B is fast.

Note that C_α^0 corresponds to pure mixing with no reaction, whereas C_α^∞ corresponds to the instantaneous reaction according to Eqs. (5.50) and (5.51). The interpolation model is now obtained by introducing a reaction progress variable Y_p ($0 \leq Y_\text{p} \leq 1$), such that the conditional concentrations are found as

$$\frac{C_\alpha}{C_{\alpha 0}} = \frac{C_\alpha^0}{C_{\alpha 0}} Y_\text{p} + \frac{C_\alpha^\infty}{C_{\alpha 0}} \left(1 - Y_\text{p}\right), \qquad \text{for} \qquad \alpha = \text{A or B}, \tag{5.68}$$

corresponding to the dotted lines in Figure 5.14. Hence the instantaneous limit is obtained for $Y_\text{p} = 0$, whereas the no-reaction limit is obtained for $Y_\text{p} = 1$. Equation (5.68) replaces Eqs. (5.50) and (5.51) in the interpolation model. However, Eq. (5.68) is still not closed due to Y_p. The closure for the reaction-progress variable reads

$$Y_\text{p} = \frac{\langle C_\alpha \rangle - \langle C_\alpha^\infty \rangle}{\langle C_\alpha^0 \rangle - \langle C_\alpha^\infty \rangle}, \qquad \text{for} \qquad \alpha = \text{A or B}. \tag{5.69}$$

The terms in angle brackets are averages computed straightforwardly as e.g.

$$\langle C_\alpha^\infty \rangle = \int_0^1 C_\alpha^\infty(\eta)\varphi(\eta)\text{d}\eta. \tag{5.70}$$

Note that Y_p is computed for just one of the species A and B, since, given either, the other can be found through Eq. (5.47). Naturally, one reaction-progress variable suffices to describe one reaction. To close Eq. (5.70) a separate transport equation for $\langle C_\alpha \rangle$ is also required:

$$\frac{\partial \langle C_\alpha \rangle}{\partial t} + \langle U_j \rangle \frac{\partial \langle C_\alpha \rangle}{\partial x_j} = \frac{\partial}{\partial x_j}\left(D_\text{T} \frac{\partial \langle C_\alpha \rangle}{\partial x_j}\right) - \gamma_\alpha k_1 \langle C_\text{A} C_\text{B} \rangle. \tag{5.71}$$

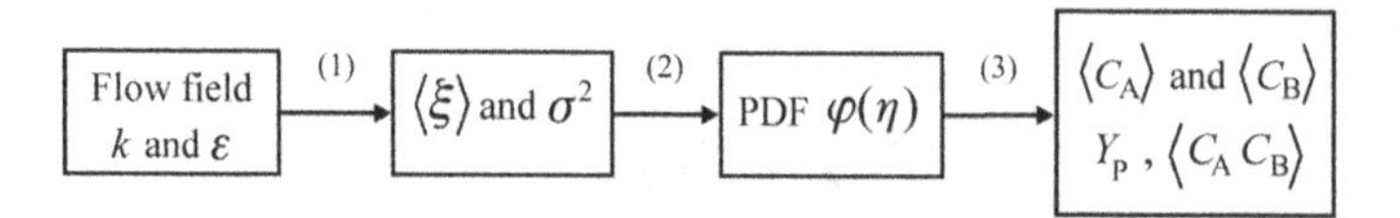

Figure 5.15 An implementation strategy for the interpolation model.

For the current purpose α can be either A or B. The average reaction rate can be computed as

$$k_1\langle C_A C_B\rangle = k_1 \int_0^1 C_A(\eta) C_B(\eta) \varphi(\eta) \mathrm{d}\eta. \tag{5.72}$$

where $C_A(\eta)$ is computed from Eq. (5.68) and $C_B(\eta)$ can be computed from Eq. (5.47) with $\xi = \eta$ and $\gamma_A = \gamma_B$ as

$$C_B(\eta) = C_A(\eta) + C_{B0} - \eta(C_{A0} + C_{B0}). \tag{5.73}$$

Equation (5.72) closes the problem and the interpolation model can be implemented using the strategy of Figure 5.15. First (1) the flow field needs to be computed to provide the turbulent kinetic energy and energy-dissipation rates. These parameters are then used for calculating the mean and variance of a mixture fraction using e.g. the TMM model (2) that further can be used to close the mixture fraction PDF from e.g. Eqs (5.13)–(5.15). With the PDF we can (3) compute the average reaction rate that closes the transport equation for the average reactive species.

Any model can be used for steps (1) and (2) in Figure 5.15 as long as the mixture-fraction PDF is delivered to the final step. The final step requires solution of Eq. (5.71) using Eqs. (5.47), (5.68)–(5.70) and (5.72). The average concentration in the computational cell is iterated. An initial guess of $\langle C_A\rangle$ gives the progress variable in Eq. (5.69) that is used for calculating the reaction rate in Eq. (5.72). This reaction rate is then used in solving Eq. (5.71), which produces a new value of $\langle C_A\rangle$. Note that there is only one more transport equation that needs to be solved in the CFD problem. All the remaining equations are algebraic and must be solved for in user-defined functions to provide the reaction rate as input for the transported species.

A more general procedure for implementing reaction-progress variables that can be more easily extended to multiple reactions has been described by Fox [15].

The state of the art for presumed PDF methods is the conditional moment closure that actually solves transport equations for the conditional concentrations in mixture-fraction space, instead of relying on the linear interpolations. The conditional moment closure has been applied both to combustion and to mixing-sensitive liquid reactions. One major disadvantage of the conditional moment closure is that the governing transport equations are five-dimensional (three spatial, one time and one mixture-fraction space) and thus computationally demanding. The large-dimensionality space is also one of the major problems of today's state of the art in modelling of reactive mixing, the full PDF methods, which solve transport equations for the joint PDF of reactive species. In full PDF methods the dimensional space is proportional to the number of reacting species, which can be substantial.

Table 5.2 Mixing and reaction of species A and B from different initial conditions and volume fractions

Scenario (a)	
$C_A = 1$	$C_A = 0$
$C_B = 0$	$C_B = 1$
$V_A = 0.5$	$V_A = 0.5$
Scenario (b)	
$C_A = 2/3$	$C_A = 0$
$C_B = 0$	$C_B = 2$
$V_A = 3/4$	$V_B = 1/4$

5.6 Non-PDF models

The most common reaction models found in most CFD software are based on eddy-dissipation (ED) modelling and do not incorporate a mixture-fraction PDF. The idea behind ED models is that the rate of mixing limits the mean rate of reaction. In ED two rates are computed: the reaction rate based on the mean concentrations (the slow-chemistry limit) and a scalar mixing time. Since the ED model was originally developed for combustion, the mixing time has traditionally been computed with the inertial–convective τ_{IC} timescale. However, better results could probably be obtained for liquids using, for example, the mean scalar dissipation rate in the turbulent-mixer model (Eq. (5.35)) or some other model developed for high Schmidt numbers. The mean reaction rate calculated with the ED model is

$$\langle S_\alpha \rangle = \min[S_{\text{kinetic}}, S_{\text{mixing}}] = \min[S(\langle \mathbf{C} \rangle),\ c_i \langle C_\alpha \rangle 1/\tau], \tag{5.74}$$

where $\langle C_\alpha \rangle$ is the limiting reactant. The rate expression means that the reaction rate cannot be faster than a constant, c_i, times the rate of mixing. In its simplest form the rate of scalar dissipation is calculated from the inertial–convective timescale $\langle N \rangle = 1/\tau \propto \varepsilon/k$. The ED model works well for slow reactions and instantaneous reactions, and is here a simple alternative to the mixture-fraction approach. It works well for certain combustion units since the combustion reaction is slow before ignition, but can be extremely fast once ignition has started. However, the ED model gives a very simplified picture of the physics and, unfortunately, for intermediate *Da* it is not possible to correlate the average reaction rate with the slow and instantaneous limits. Further, the model cannot predict the dependence of mixing-sensitive reactions on the initial volume fractions of the reacting flows. From experiments it is well known that the reaction rates depend very much on the initial volume fractions of the fluids and the two scenarios in Table 5.2 should lead to very different average reaction rates despite the fact that the average concentrations are equal, $\langle C_A \rangle = \langle C_B \rangle = 0.5$.

This dependence can be captured only by incorporating the structure of the mixture, which is achieved with the more advanced mixture-fraction models discussed above. The ED model can be useful for predicting trends for instantaneous reactions, and is useful if there are multiple injections, recirculation streams or other complicating factors.

The ED model is applicable for any initial configuration, which mainly accounts for its popularity in commercial software.

5.7 Summary

In this chapter we have discussed the problems specific to reactive mixing of turbulent incompressible, isothermal flows. The major emphasis has been on modelling of a conserved scalar and the use of a conserved scalar in modelling of chemical reactions, through presumed PDF methods.

Questions

(1) What is the most important parameter to study when you are first presented with a reactive-mixing problem?
(2) What is meant by the mixture-fraction PDF? What is a conserved scalar?
(3) What is the physical interpretation of the mixture-fraction variance?
(4) Explain the most important features of the turbulent-mixer model.
(5) What is the smallest relevant length scale for turbulent mixing? What is the physical interpretation of this scale?
(6) What is described by the Schmidt number? What is described by the turbulent Schmidt number?
(7) Why is it difficult to solve problems in which $Da \approx 1$?

6 Multiphase flow modelling

Most books on fluid mechanics, especially books on turbulence, will have a statement along the lines of 'most important fluid flows are turbulent'. That statement can be made for multiphase flows as well. Most of the flows around us are multiphase. They exist in nature, such as rain or snow in the sky, or the flow of a river, which may transport all kinds of solids and has a large interface with the air above it. Also, the flow of blood in our veins is a good example of a mixture of a fluid and particles. Such cases prevail even more so in industry, especially in the processing industry. In the pharmaceutical industry, pills are made from small particles. Also droplets can be dispersed so that they can be inhaled into lungs. There is an abundance of multiphase flows in chemical-process industries, e.g. flow of catalyst particles, fuel, plastics, gases, etc. The energy-producing industry has many examples too, such as the burning of coal particles and fuel sprays and the boiling of water. Bubbly flows prevail in the nuclear industry, where the science of studying multiphase flows all started. These are just a few of the multitude of examples of multiphase flows. Some examples of multiphase flows and their applications are shown in Table 6.1.

The word 'phase' in multiphase flows refers to the solid, liquid or vapour state of matter. The prefix 'multi' means multiple. So a multiphase flow is the flow of a mixture of phases such as gases (bubbles) in a liquid, or liquid (droplets) in gases or particles in liquids and/or gases. This definition should not be taken too strictly; for instance, the flow of two immiscible liquids does not contain multiple phases in a thermodynamic sense, yet, because there are multiple different liquids, and this is in fact quite similar to the situation of a droplet in a gas flow, they are still considered within the research area of multiphase flow. A more detailed disussion of multiphase flows can be found in [16].

The research area of multiphase flows is extremely broad and not very well defined. This has led to a very wide field of research, both fundamental and applied. It has also led to a lot of confusion – so much confusion that, today, there is not even agreement upon the governing equations which are to be used, let alone all the empirical closure models obtained from measured data. There are some areas where reliable ab-initio simulations are possible, whereas in other areas only parameter studies around experimentally validated simulations are possible. In general simulations of multiphase flows are reliable at low particle loading with particles that follow the continous phase closely. In contrast, multiphase systems that are dominated by non-ideal particle–particle collisions are very difficult so simulate accurately. This chapter tries to give an overview of the modelling

Table 6.1 Summary of two-phase flow systems and some industrial applications

Continuous phase–dispersed phase	Industrial applications
Gas–solid flows	Pneumatic conveying, fluidized beds, solid separation (filters, cyclones)
Liquid–solid flows	Stirred vessels, liquid–solid separation, hydraulic conveying
Gas–liquid (droplet) flows	Spray drying, spray cooling, spray painting
Liquid–droplet flows	Mixing, separations, extraction
Liquid–gas (bubble) flows	Flotation, aeration, bubble columns

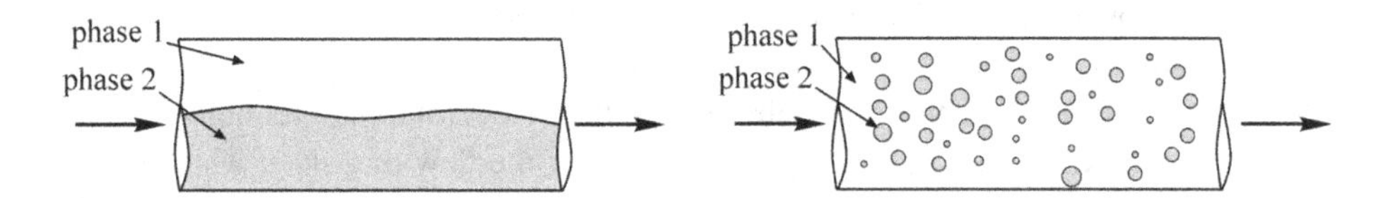

Figure 6.1 A separated (stratified) multiphase flow (left) versus a dispersed multiphase flow (right).

possibilities for dispersed and separated two-phase flows. Bubble and drop break-up and coalescence and population-balance modelling are not included.

6.1 Introduction

An important classification of multiphase flow is made in terms of whether the different phases present in the flow are separated or dispersed, as shown in Figure 6.1. In a dispersed flow, one phase is typically present in the form of particles or droplets and there are many individual interfaces. In a separated flow, the phases present are relatively separated, with only a few interfaces.

Commonly, dispersed two-phase flows are separated into two types of flow regime, the dilute regime and the dense regime. In the dilute regime, the spacing between the particles or droplets is quite large, so their behaviour is governed by the continuous phase (fluid) forces. In dense phase systems, the spacing is smaller, so the inter-particle interactions are typically very important. Very roughly, flows with a spacing of less than ten particle diameters are considered to be dense.

6.1.1 Characterization of multiphase flows

Several, mostly dimensionless, parameters are used to characterize multiphase flows. The most important one is the volume fraction, which defines how much of the local volume is occupied by either of the phases. The dispersed-phase volume fraction is the volume occupied by the particles in a unit volume,

$$\alpha_{\mathrm{d}} = \frac{\sum_{i=1}^{N_{\mathrm{d}}} V^i}{V}, \tag{6.1}$$

where V^i is the volume occupied by particle or droplet i and N_d is the total amount of particles or droplets present in volume V. The characteristic size of the dispersed phase is given as

$$D_{d,i} = \frac{6(V^i)^{1/3}}{\pi}, \tag{6.2}$$

which assumes that the volume occupied by one particle has a spherical shape. For some types of particle this is not a very good approximation, for instance for a fiber or a deformed droplet. Then, a sphericity factor depending upon the area or volume of the particle as seen by the flow can be introduced.

The 'typical' distance between the particles, assuming that they are homogeneously arranged, is given by

$$L = D_d \left(\frac{\pi}{6\alpha_d} \right)^{1/3}, \tag{6.3}$$

and this is one of the parameters used to determine the importance of inter-particle interactions. The volume fraction of the continuous phase is given as

$$\alpha_f = 1 - \alpha_d. \tag{6.4}$$

Typically, the inertia of the phases in multiphase flow modelling is expressed in terms of the bulk density, $\alpha_d \rho_d$, instead of just the intrinsic density ρ_d. The bulk density is a measure for the potential inertia, and will thus be employed to construct the mass and momentum balances. The mixture density is the sum of the dispersed-phase density and the continuous-phase density.

Sometimes, in the engineering literature, the flux of the dispersed phase is expressed in terms of the mass loading, which is defined as the mass flux of the dispersed phase divided by that of the continuous phase,

$$m = \frac{\alpha_d \rho_d U_d}{(1 - \alpha_d) \rho_f U_f}. \tag{6.5}$$

Timescales and length scales are important measures in fluid mechanics and even more so in the case of the timescale for a particular physical mechanism in the multiphase flow. Examples of these mechanisms are collisions, inertia and dissipation.

It is difficult to define one unique scale for a flow or a dispersed phase, since these are given by large distributions. But it is still useful to determine or estimate a dominant dimensionless number specifying the ratio of a dispersed-phase timescale and a continuous-phase timescale. This ratio is called the Stokes number, and a Stokes number much bigger than unity means the particles are relatively insensitive to that specific timescale of the continuous-phase behaviour. For instance, the turbulence Stokes number is given as

$$St_T = \frac{\tau_d}{\tau_T}, \tag{6.6}$$

where τ_d is the timescale of the dispersed phase and τ_T is the relevant timescale of the turbulence. An estimation of the particle response time is obtained by solving Eq. (6.9)

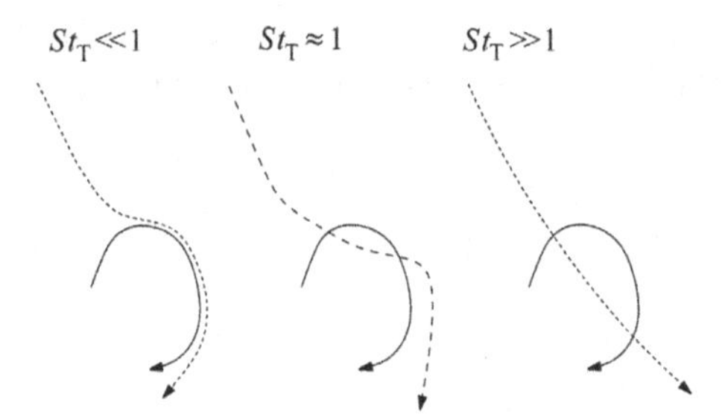

Figure 6.2 Effects of a turbulent eddy (solid line) on particle trajectory (dashed line) for different Stokes-number limits.

below. For simple flow at low Reynolds numbers the response time will be

$$\tau_d = \frac{\rho_d D^2}{18\mu_f}. \tag{6.7}$$

When $St_T \to 0$, for instance for very small particles, the particles follow the flow completely. For flows in which $St_T \to \infty$, the behaviour of the particles is uncorrelated with the flow characteristics. This effect is depicted in Figure 6.2.

Another important Stokes number is the collision Stokes number,

$$St_c = \frac{\tau_d}{\tau_c}, \tag{6.8}$$

where τ_c represents the timescale of the inter-particle interactions. If $St_c < 1$, a flow may be assumed dilute; if $St_c > 1$, a flow may be considered dense. This measure can be used as an additional measure to determine the importance of particle–particle interactions, next to that of particle spacing.

6.1.2 Coupling between a continuous phase and a dispersed phase

A general classification of dispersed two-phase flows with regard to the interaction with the continuous phase was provided by Elgobashi in 1994. There are three possible ways of coupling, which are listed in Table 6.2. If the particles do not influence the flow and their interactions are negligible, only the effect of the fluid on the particles is important. The terms describing how the particles affect the flow and how they affect each other may be neglected. This is defined as one-way coupling.

If particles have a Stokes number larger than unity or the volume fraction is sufficiently large to affect the average denisty of the mixture, their effect on the flow is no longer negligible, and modelling of the effect of the particles on the flow has to be incorporated into the governing equations for the continuous phase. If the flow is still sufficiently dilute, inter-particle interactions may safely be neglected. This is defined as two-way coupling.

If the volume fraction of the dispersed phase is sufficiently large, typically above 10^{-3}, inter-particle interactions become important. Inter-particle interactions may be collisions, but also more indirect phenomena, such as two particles approaching each

Table 6.2 The various types of coupling between the dispersed phase and the continuous phase. The choice of coupling can be made on the basis of the length scale of the dispersed phase, or on the basis of the Stokes number

Type of coupling	Definition	Implementation
One-way	Dispersed phase senses continuous phase, but the continuous phase is unaffected	Particle tracking may be done in a post-processor
Two-way	Dispersed phase senses continuous phase and continuous phase senses dispersed phase	The presence of the dispersed phase should be reflected in the governing equations of the fluid
Four-way	Dispersed particles interact	A model for particle–particle interactions should be included

other in a viscous liquid. Although they will most probably not collide, their interaction may still be important. This is defined as four-way coupling (note that one particle affecting the other and the other affecting the first is counted as two coupling terms).

Note that this classification does not specify how accurate a model should be. The dispersed phase may be represented by a momentum source in the governing equations for the continuous phase, or, alternatively, the complete flow around a particle may be resolved. Although they are very different in terms of the level of detail, both these models fall into the category of two-way coupling.

6.2 Forces on dispersed particles

Newton's second law can be applied to the dispersed particles. The main controversy has been over how to describe the forces acting on the particles. The original work by Basset (1888), Boussinesq (1885) and Oseen (1927), abbreviated as the BBO equation, described a few small particles in a uniform flow with particle Reynolds number $Re_\mathrm{P} \ll 1$. The original work was later adjusted for higher particle Reynolds numbers and turbulent flows. In a more general framework, the forces on a single particle are given by

$$m_\mathrm{d}\frac{\mathrm{d}U_{i,\mathrm{d}}}{\mathrm{d}t} = F_{i,\mathrm{Drag}} + F_{i,\mathrm{Press}} + F_{i,\mathrm{Virt}} + F_{i,\mathrm{History}} + F_{i,\mathrm{Bouy}} + F_{i,\mathrm{Lift}} + F_{i,\mathrm{Therm}} + F_{i,\mathrm{Turb}} + F_{i,\mathrm{Brown}}, \tag{6.9}$$

where $U_{i,\mathrm{d}}$ is the linear velocity of the particle, m_d is the mass of the particle, $F_{i,\mathrm{Drag}}$ is the drag force, $F_{i,\mathrm{Press}}$ is the pressure force due to the pressure gradient, $F_{i,\mathrm{Virt}}$ is the virtual mass force due to acceleration of the surrounding fluid, $F_{i,\mathrm{History}}$ is the history or Basset force due to changes in the boundary layer, $F_{i,\mathrm{Bouy}}$ denotes forces due to gravity, $F_{i,\mathrm{Lift}}$ is the Saffman and Magnus lift force due to the velocity gradient and particle rotation, $F_{i,\mathrm{Therm}}$ is the thermophoretic force due to a temperature gradient, $F_{i,\mathrm{Turb}}$ denotes forces due to turbulent fluctuations and $F_{i,\mathrm{Brown}}$ is the Brownian force due to molecular collisions. These forces are described briefly below and a more detailed discussion can be found in [17].

The Magnus force is due to the rotation of the particle and a torque balance is required. This balance between rotational velocity, ω_i, inertia, I_i, and torque, T_i, is given by

$$I_i \frac{\mathrm{d}\omega_i}{\mathrm{d}t} = T_i.$$

However, the Magnus force is usually small and not well defined outside the laminar region, and the torque balance is not discussed further in this book.

The forces in Eq. (6.9) above are briefly described below. Note that the detailed flow around the particles is not resolved. The model requires that the particles are much smaller than the computational cell and that the continuous phase is resolved on a large scale. The effect of the dispersed phase on the continuous phase is treated as a source term arriving from all particles in the cell.

For a single particle the *drag force* is expressed as a function of the relative velocity between the two phases,

$$F_{i,\mathrm{Drag}} = \frac{1}{2} A_\mathrm{d} C_\mathrm{D} \rho_\mathrm{f} \left| U_\mathrm{f} - U_\mathrm{d} \right| \left(U_{i,\mathrm{f}} - U_{i,\mathrm{d}} \right), \tag{6.10}$$

where A_d is the projected area normal to the flow, i.e. $\pi D_\mathrm{p}^2/4$ for a sphere. The drag coefficient C_D is discussed in more detail in Section 6.4.1.

The second term represents the *pressure and shear forces* from the fluid on the particle. This is usually expressed in terms of the pressure and shear gradient over the particle surface. Assuming a constant pressure and shear gradient over the volume of the particle, this force can be written

$$F_{i,\mathrm{Press}} = V_\mathrm{d} \left(-\frac{\partial P}{\partial x_i} + \frac{\partial \tau_{ij}}{\partial x_j} \right), \tag{6.11}$$

where V_d is the volume of the particle.

The third term represents the *virtual-*, *apparent-* or *added-mass force*. This force arises from the acceleration or deceleration of the fluid surrounding an accelerating or decelerating particle. The effect of this term is an increase in the apparent mass of the particle, whence the name added-mass force. This force is written in the form

$$F_{i,\mathrm{Virt}} = -C_\mathrm{VM} \rho_\mathrm{f} V_\mathrm{d} \frac{\mathrm{D}}{\mathrm{D}t} (U_{i,\mathrm{d}} - U_{i,\mathrm{f}}), \tag{6.12}$$

where $\mathrm{D}/\mathrm{D}t$ is the substantial operator and represents the relative acceleration of the particle compared with the fluid along the path of the particle. The virtual-mass force coefficient C_VM is usually close to 0.5, which indicates that a volume of the continuous phase corresponding to half the volume of the particle is accelerated with the particle. This force can be neglected when the density of the continuous phase is much lower than the density of the particle and the virtual mass is much less than the mass of the particle.

The fourth term, the *history force*, arises from the time required to develop the boundary layer around the particle when the particle is accelerated or decelerated. This development leads to a separation of timescales between the fluid and the particle, thereby creating the necessity for the time integral in the force. This time integral makes the history force computationally very expensive – the calculation times may increase by an order of magnitude.

The fifth term represents the *bouyancy force*, the volume of the particle multiplied by the density difference between the phases and the gravitational acceleration constant.

The *Saffman and Magnus lift forces* are due to the higher velocity on one side of the particle arising from flow in a velocity gradient (the Saffman lift force) or rotation of the particle (the Magnus lift force). Accurate models are available only for spherical bodies at low particle Reynolds numbers. There is very little empirical data for both types of lift force at higher Reynolds numbers. Owing to boundary-layer separation and deformation of fluid particles, even the direction of the lift force can be difficult to predict.

The *thermophoretic force* represents the force due to a temperature gradient in the fluid. Hot molecules move faster than cold molecules and a large temperature gradient will give a net force in the direction opposite to the temperature gradient. This thermophoretic force is important only for very small particles, and will lead to a separation of particles depending on their size.

The origin of the *Brownian force* is random collisions of individual molecules. This force is usually modelled as Gaussian white noise. The momentum transferred by collision of individual molecules is very small and the Brownian force is important only for submicrometer particles.

The *forces due to turbulence* are often modelled as a random addition to the fluid velocity that is sustained for a time corresponding to the minimum of the lifetime of the turbulent eddies and the time taken for a particle to pass through a turbulent eddy.

The importance of the terms in the above equation of motion for one particle can be analysed by dividing all the forces by the particle density. If the particle density is much larger than the fluid density, as in gas–solid and gas–droplet flows, terms linear in ρ_f/ρ_d may be neglected. This means that only drag, the pressure and shear-stress gradient, and the buoyancy are important in such flows. If terms containing ρ_f/ρ_d may not be neglected, such as for rising bubbles and liquid–solid flows, the added-mass force and the history force are typically important. It is much more difficult to calculate the trajectories of particles in such flows.

The original BBO equation is valid for one (or very few) particles in a homogeneous flow with particle $Re < 1$. Most industrial applications do not deal with a few particles in a homogeneous and low-Reynolds-number flow, so empirical extensions of the forces from their original derivation are required. For the drag force, this has been a very successful approach; here the empirical coefficient is called the drag coefficient. The lift coefficient, virtual-mass coefficient, and history-force coefficient have been found to be more prone to error, and the form of the terms in Eq. (6.9) may be less suitable. Closure models for these terms are discussed later.

6.3 Computational models

There are many different kinds of model available for multiphase flow. The models presented in this book can be subdivided into five main classes:

- the Euler–Lagrange model
- the Euler–Euler model
- the mixture or algebraic-slip model
- the volume-of-fluid (VOF) model
- porous-bed models.

In *Euler–Lagrange modelling* the fluid phase is modelled as a continuum by solving the Navier–Stokes equations, while for the dispersed phase a large number of individual particles is modelled. The dispersed phase can exchange momentum, mass and energy with the fluid phase. Since the particle or droplet trajectories are computed for each particle or for a bundle of particles that are assumed to follow the same trajectory, the approach is limited to systems with a low volume fraction of dispersed phase.

In *Euler–Euler models* the different phases are all treated as continuous phases, and momentum and continuity equations are solved for each phase. The Euler–Euler model can handle very complex flows, but does not always give the best results since empirical information is needed in order to close the momentum equations. Typical applications are risers and fluidized beds.

In the *mixture model* (algebraic-slip model) the flows of phases are assumed to interact strongly and it is not necessary to solve the momentum balances for the different phases separately. In this model the viscosity is estimated for the mixture. The velocities of the different phases are thereafter calculated from buoyancy, drag and other forces, giving the relative velocities in comparison with the mean velocity of the mixture. Typical applications are bubble columns, fine particle suspensions and stirred-tank reactors.

The *volume-of-fluid (VOF) model* is an Euler–Euler model whereby the interface between the different phases is tracked. The model is suitable for stratified flow, free surface flows and movement of large bubbles in liquids. Since the interface between the fluids must be resolved, it is not applicable for a system with many small drops or bubbles.

In the *porous-bed model*, the pressure drop across a porous bed is modelled. In a bed containing many particles, it is not possible to resolve the geometry and solve the Navier–Stokes equations. Instead the pressure drop is calculated from an equation similar to the Ergun equation for the pressure drop in fixed beds [18].

6.3.1 Choosing a multiphase model

In selecting the most appropriate multiphase model, the physics of the system must be analysed and understood. Initially, there are some questions that must be asked.

- Are the phases separated or dispersed?
- Will the particles follow the continuous phase? What is the Stokes number?
- How large are the local volume fractions?
- How many particles are there in the system?
- What kind of coupling occurs? Is it one-, two- or four-way coupling?

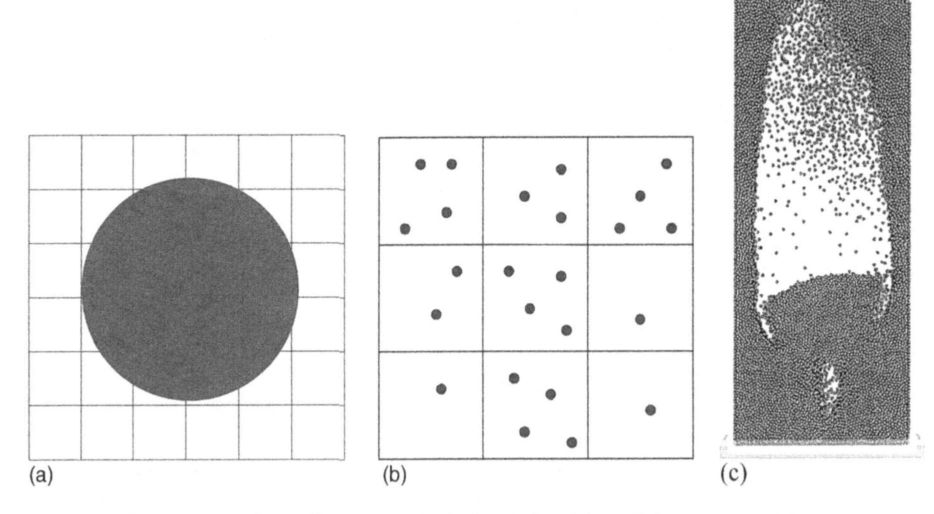

Figure 6.3 The concept of true direct numerical simulation (a), and the point-particle approach (b). In (c) an example of the point-particle approach is shown.

For separated flows only the VOF model will work. The model predicts the location of the interface and uses single-phase models to predict the flow in each phase. The model requires a fine mesh to resolve the curvature of the interface.

The porous-bed model is applicable to a system dominated by viscous and inertial forces, and the pressure drop can be calculated as a function of the flow properties by using an empirical a-priori given function.

For dispersed multiphase systems several models are possible. The most accurate is usually the Euler–Lagrange model. It works well for systems with one- or two-way coupling, but requires additional closures for four-way coupling. In such cases the computational time increases and the quality of the simulations is poor except for very ideal systems. The limitation for the Euler–Lagrange model is the number of particles. A few hundred thousand particles or bundles of particles is the limit on a desktop computer.

The general models for dispersed multiphase flow are the mixture and Euler–Euler models. The mixture model requires that the Stokes number is low and that the phases accelerate together. The mixture model is more stable and faster than the Euler–Euler model and should be used whenever possible. It may also be used to obtain good initial conditions for an Euler–Euler simulation. The Euler–Euler approach can be used when no other model is possible.

6.3.2 Direct numerical simulations

Resolving the behaviour of a flow all the way down to the smallest scales is called direct numerical simulation (DNS). If particles have a fluid Stokes number larger than unity, hence two-way coupling, the flow around the particles must be resolved as shown in Figure 6.3(a). Although this is possible, it requires efficiently developed algorithms and

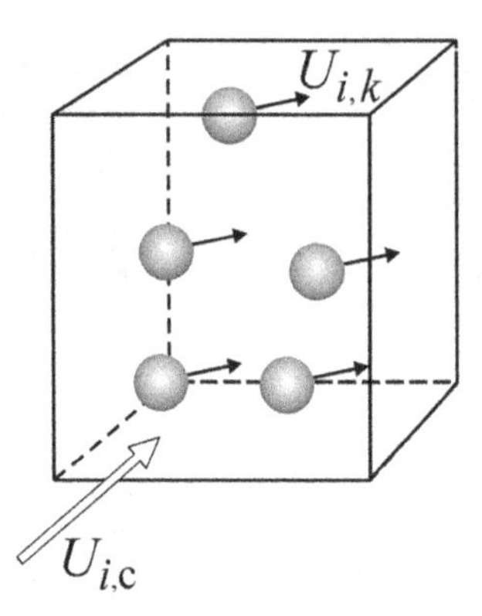

Figure 6.4 Lagrangian control volume.

a lot of patience; at the time of writing, only about 10^3 particles can be considered. This might be useful for research objectives, but is useless for industrial purposes.

6.3.3 Lagrangian particle simulations, the point-particle approach

In this technique the particles are tracked individually, and the gas phase is treated in a continuous framework. This can be done by resolving the flow around the particles, or by representing the particles as source terms in the flow. This is depicted in Figures 6.3(a) and (b), and in Figure 6.3(c) a Lagrangian particle simulation for a small fluidized bed is shown.

In point-particle-approach simulations, the single-phase Navier–Stokes equations for the continous phase are solved in conjunction with tracking the individual particles,

$$\frac{\partial(\alpha_f \rho_f)}{\partial t} + \frac{\partial(\alpha_f \rho_f U_{i,f})}{\partial x_i} = S_C, \tag{6.13}$$

$$\frac{\partial(\alpha_f \rho_f U_{i,f})}{\partial t} + U_{i,f}\frac{\partial(\alpha_f \rho_f U_{j,f})}{\partial x_i} = -\alpha_f \frac{\partial P}{\partial x_i} + \frac{\partial(\alpha_f \tau_{ij,f})}{\partial x_j} + S_{i,p}, \tag{6.14}$$

where S_C is a source term describing mass transfer between the phases and $S_{i,p}$ represents momentum exchange between the particles and the fluid. All forces on the right-hand side of Eq. (6.9) except gravity are due to interaction with the continous phase and must appear in the source term $S_{i,p}$ of Eq. (6.14). Equation (6.14) is written per volume and Eq. (6.9) per particle, and the term $S_{i,p}$ must include the number of particles, per volume, i.e. $n/V = 6\alpha_d/(\pi D_p^3)$ assuming spherical particles. For successful employment of the Euler–Lagrange model, the particles have to be much smaller than the fluid-phase grid cells as shown in Figure 6.4. This restriction arises because the velocity field, U_f, required to calculate the source term needs to be the undisturbed velocity field.

The flow of the continuous fluid (Eqs. (6.13) and (6.14)) can be solved with traditional RANS or LES models with the additional terms describing the interaction between the continuous and dispersed phases. The movement of all the particles is simulated by integrating the trajectory Eq. (6.15) and the force balance Eq. (6.9) with given initial position for all particles:

$$\frac{dx_i}{dt} = U_{i,\mathrm{d}}. \tag{6.15}$$

The number of particles is limited because it involves solving an ODE for all particles. However, it is possible to bundle particles that behave identically into packages containing thousands of particles. This will give a correct source term for the continuous phase. The limitation is that the bundle will be modelled assuming that the properties at the centre of gravity for the bundle are valid for all particles and that the source term for the bundle is at the centre of gravity.

Euler–Lagrange models are usually accurate at low volume fraction with one- or two-way coupling. At higher volume fraction, when the particles collide the model requires additional closures (see Section 6.4.2). The simulations become very demanding at high particle loading due to the high number of collisions. It is not possible to calculate all potential collisions beween all particles, and most CFD programs simulate collisions only for particles that are within the same computational cell. More advanced algorithms may also include neighbouring cells. In all cases, the number of particles must be low and the time step must be limited so that no particle moves by more than one computational cell in one time step. In addition it is not possible to model how the particles will collide. Even if the momentum is conserved and the absolute value of the velocity is known, the direction is unknown. There are stochastic models that calculate a probability distribution of velocities of a large number of collisions after each time step. However, the use of Euler–Lagrange models with four-way coupling is not yet a feasible approach in engineering.

Turbulence modelling

The continuous phase may be modelled using standard RANS or LES methods. In the k–ε model a source term for the additional turbulence energy arising from the movement of the particles may be included. The turbulence energy generated from the movement of particles can be formulated as a source term in the equation for k:

$$S_k = \frac{\alpha_\mathrm{d}\rho_\mathrm{d}}{\tau_\mathrm{d}\rho_\mathrm{c}} \left(U_\mathrm{d} - \langle U_\mathrm{f}\rangle\right)^2 . \tag{6.16}$$

The dissipation is assumed to increase in proportion to the increase in kinetic energy divided by the timescale for the large eddies, k/ε, as in the standard ε equation

$$S_\varepsilon = C_{\varepsilon 3}\frac{\varepsilon}{k}S_k, \tag{6.17}$$

where the constant $C_{\varepsilon 3}$ is about 1.8. However, the size of the turbulent eddies formed by the particles is usually much smaller than that of the energy-containing eddies that contain most of the turbulent kinetic energy and also transport most of the momentum. The small eddies formed by the particles will decay very fast and transport only a small fraction of the momentum. The extra source term should be included only when the turbulent eddies formed by the particles are large compared with the energy-containing eddies.

The effect of fluid-phase turbulence on the particles may be important, depending upon the Stokes number. In determining a drag force, the instantaneous drag forces on each particle are averaged into a drag force, which becomes a function of the mean or average relative velocity. The turbulent eddies in the continous phase will move the single particles in all directions and away from the average path. At the end particles injected at the same point will not end up at the same position due to the turbulence, and this phenomenon must be taken into account.

The information on the turbulence provided by RANS models is available only in statistical terms, for instance, in terms of the turbulent kinetic energy and energy-dissipation rates. To couple the behaviour of the particles effectively with the local fluid properties, however, instantaneous fluid properties are required. A popular way of determining a non-unique set of instantaneous fluid properties from the given statistical turbulence models is use of the so-called random-walk models. The instantaneous fluid velocity at a particle location is determined from the averaged fluid velocity, which is deterministic, and a random component for which the magnitude, direction and timescales obey the statistical (averaged) properties of the local fluid turbulence.

In the discrete-random-walk (DRW) model it is assumed that the effect of turbulence can be modelled by adding a random velocity to the continuous phase for a specific time T, where T is the time the particle spent in a turbulent eddy. This time is estimated as the minimum of the time taken for a particle to pass through the turbulent eddy and the lifetime of the turbulent eddy. The time taken to pass through the turbulent eddy is estimated from the size of the turbulent eddy and the slip velocity,

$$T \approx -\tau \ln\left(1 - \frac{l}{\tau |U_{\mathrm{d}} - \langle U_{\mathrm{f}} \rangle|}\right), \tag{6.18}$$

where τ is the particle response time (Eq. (6.7)) and l is the size of the turbulent eddy (Eq. (4.19)). The lifetime of a turbulent eddy in RANS models is

$$T \propto \frac{k}{\varepsilon}. \tag{6.19}$$

The turbulent velocity fluctuations are assumed to have a Gaussian probability distribution and the turbulent fluctuations are added to the continuous-phase velocity

$$u_i' = \xi_i \sqrt{u_i^2}, \tag{6.20}$$

where ξ_i is a Gaussian random number.

In isotropic turbulence $\langle u_1^2 \rangle = \langle u_2^2 \rangle = \langle u_3^2 \rangle = 2k/3$ and the continuous-phase velocity fluctuation is calculated from

$$u_i = \xi_i \sqrt{\frac{2k}{3}}. \tag{6.21}$$

All three velocity fluctuations are known in Reynolds stress modelling and all the individual fluctuating components can be calculated. The random velocity u_i' is then added to the average fluid velocity $\langle U_{i,\mathrm{f}} \rangle$ and kept constant during the time step defined by T above. A new random number ξ_i is chosen after each time step.

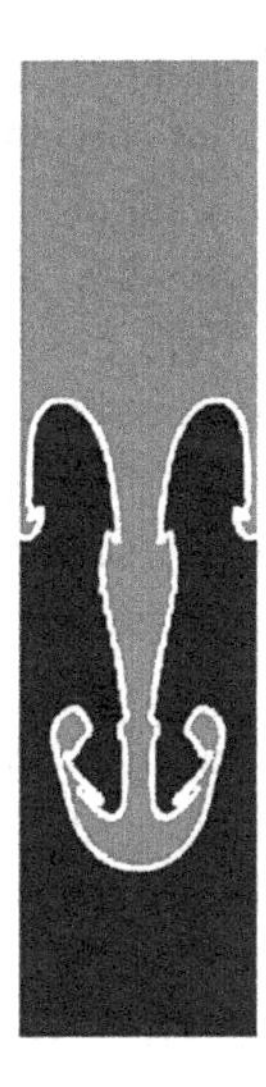

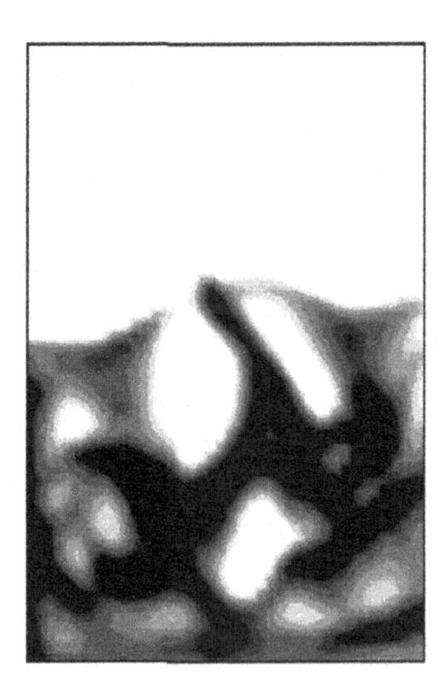

Figure 6.5 A stratified flow simulation (left) versus a dispersed flow simulation (right).

6.3.4 Euler–Euler models

The Euler–Euler model treats a dispersed multiphase flow as two (or more) fully interpenetrating quasi-fluids, and is therefore often referred to as the two-fluid model. The two-fluid model is derived by ensemble averaging or volume averaging. A very important quantity appearing in the equation because of the averaging is the volume fraction α_k. This quantity alone does not say anything about the size or behaviour of the dispersed phase, and this generally comes in via closure models. Within the Euler–Euler framework both stratified and dispersed flow are modelled. In this book we present the Euler–Euler model as a general model for most kinds of multiphase flow, the mixture model as a simplification of the Euler–Euler model and the volume-of-fluid model for stratified flows as shown in Figure 6.5.

Intuitively, it is easiest to imagine the derivation of the two-fluid model in terms of volume averaging. A small volume, much smaller than that of the large-scale flow structures, but much larger than that of individual dispersed particles, in which both phases are present is used in this model. The volume fraction is defined on the basis of the distribution of phases and the size of the computational volume. The local instantaneous equations describing both phases may then be averaged in the volume, considering the bulk density of each of the phases. The two-fluid model equations for each phase are given by the following equations, where k is not an index but represents a phase:

$$\sum_k \alpha_k = 1, \tag{6.22}$$

$$\frac{\partial \alpha_k \rho_k}{\partial t} + \frac{\partial \alpha_k \rho_k U_{i,k}}{\partial x_i} = -\sum_{l=1}^{p} (\dot{m}_{kl} - \dot{m}_{lk}), \tag{6.23}$$

$$\frac{\partial \alpha_k \rho_k U_{i,k}}{\partial t} + \frac{\partial \alpha_k \rho_k U_{i,k} U_{j,k}}{\partial x_j} = -\alpha_k \frac{\partial P}{\partial x_i} + \frac{\partial \alpha_k \tau_{ij,k}}{\partial x_k} + \alpha_k \rho_k g_i + F_{i,k}, \tag{6.24}$$

where p represents the number of phases, $\dot{m}_{kl}$ mass transport from phase k to phase l and F_k the interaction force with the other phases. The difficulty of solving these sets of equations is twofold. First, these equations are difficult from a numerical point of view, since there are many coupled equations with one shared pressure. Secondly, to solve the equations, closure models are required for τ and F_k. Here τ represents the rheology of the phase, and very complex models are typically required if the phase is not a Newtonian fluid but, for instance, a particle mixture. Models for estimation of viscosity in multiphase flows are presented in Section 6.4.2. The interaction force with the other phases, F_k, typically comprises collisions with the other dispersed phases and all important physical mechanisms described by Eq. (6.9).

It is important to realize that all variables appearing in this equation are *averaged* variables, rather than real point variables. This makes concepts such as LES and RANS for the two-fluid model quite complex; the equation does not revert back to a DNS simulation on reducing the grid size.

All physical phenomena, apart from the quantities provided by the principle of momentum conservation, must be modelled by closure models. This includes the rheology of the dispersed phase and the momentum transfer between the phases. The continuous fluid is often modelled using a k–ε or RSM model, but the dispersed phases need more elaborate models. Some of these closure models are discussed later in this chapter.

Turbulence modelling

Standard k–ε and RSM models can be used with the Euler–Euler multiphase model for dilute systems and when the phases can be approximated with one set of momentum models for the mixture as in Section 6.3.2. For the continuous phase in a dilute system k is modelled with the standard k equation with an additional source term describing the additional turbulence energy arising from the relative velocities of the continuous phase and the dispersed phases. For the dispersed phases the timescales and length scales for the particles are used to evaluate dispersion coefficients and the turbulent kinetic energy for each phase.

For dense systems, when a turbulence model is required for each phase, the commercial CFD software usually includes only the k–ε model. These models tend to be very unstable, and the quality of the simulations is usually low. The simulations often need calibration and should be combined with validation experiments in similar systems.

6.3.5 The mixture model

The mixture model is similar to the Euler–Euler model, but assumes one more simplification. This simplification is that the coupling between the phases is very strong and the relative velocity between the phases is in local equilibrium, i.e. they should accelerate together. In performing a simulation with the mixture model, one set of equations is solved for the mixture, i.e. the unknowns are the flow properties of the mixture, not those of the individual phases. The flow properties of the individual phases can be reconstructed with an algebraic model for the relative velocity, which is often referred to as the algebraic-slip model. The individual phase's velocity relative to the mean velocity is called the drift velocity and is denoted $U_{i,\text{dr},k}$ for phase k, while the velocity relative to the continuous phase is the slip velocity. The advantage of employing the mixture model is that only one set of equations is computed, leading to a substantial decrease of computational effort compared with the full Euler model. This set of equations is given as

$$\frac{\partial \rho_\text{m}}{\partial t} + \frac{\partial(\rho_\text{m} U_{i,\text{m}})}{\partial x_i} = 0, \tag{6.25}$$

$$\frac{\partial(\rho_\text{m} U_{i,\text{m}})}{\partial t} + \frac{\rho_\text{m}\,\partial(U_{j,\text{m}} U_{i,\text{m}})}{\partial x_j} = -\frac{\partial P}{\partial x_i} + \frac{\partial \tau_{ij,\text{m}}}{\partial x_j} + \rho_\text{m} g_i - \frac{\partial \sum_k \alpha_k \rho_k U_{i,\text{dr},k} U_{j,\text{dr},k}}{\partial x_j}, \tag{6.26}$$

where the subscript m represents the mixture property and $U_{i,\text{dr},k}$ the drift velocity for phase k, so that

$$U_{i,k} = U_{i,\text{dr},k} + U_{i,\text{m}}. \tag{6.27}$$

The last term in Eq. (6.26) arises from the nonlinear inertial term in the Navier–Stokes equations, which can be written as the second term on the left-hand side plus the last term on the right-hand side in Eq. (6.26), since

$$\frac{\partial \sum\limits_k \alpha_k \rho_k U_{i,k} U_{j,k}}{\partial x_j} = \frac{\rho_\text{m}\,\partial(U_{i,\text{m}} U_{j,\text{m}})}{\partial x_j} + \frac{\partial \sum\limits_k \alpha_k \rho_k U_{i,\text{dr},k} U_{j,\text{dr},k}}{\partial x_j}. \tag{6.28}$$

The mixture properties are typically weighed by the volume fraction, e.g.

$$\mu_\text{m} = \sum_\text{m} \alpha_k \mu_k, \tag{6.29}$$

where the viscosity μ_k for dispersed flows may be estimated using the same models as in standard Euler–Euler modelling (see the discussion of granular flow models in Section 6.4.2).

Typically, a steady-state algebraic expression based on Eq. (6.9) is specified for the drift or slip velocity for each phase to close the mixture model. The volume fractions can be determined from the conservation equation for the continuity of each phase as described by Eq. (6.23).

Turbulence modelling

Standard RANS and LES models can be used for turbulence modelling using the average properties for density and viscosity. In the k–ε model the turbulent viscosity is calculated from k and ε in the same way as for single-phase flow with a small correction for systems in which the drift velocity is large compared with the velocity of the turbulent eddies. The effect of turbulent diffusion is also added to the drift velocity calculated using Eq. (6.9),

$$(U_{i,\text{dr},k})_\text{turb} = U_{i,\text{dr},k} - \frac{\mu_\text{t}}{\sigma_\text{t}}\left(\frac{1}{\alpha_k}\frac{\partial \alpha_k}{\partial x_i} - \frac{\partial \alpha_\text{f}}{\partial x_i}\right), \tag{6.30}$$

where the turbulence Schmidt number, σ_t, is of the order of 0.7.

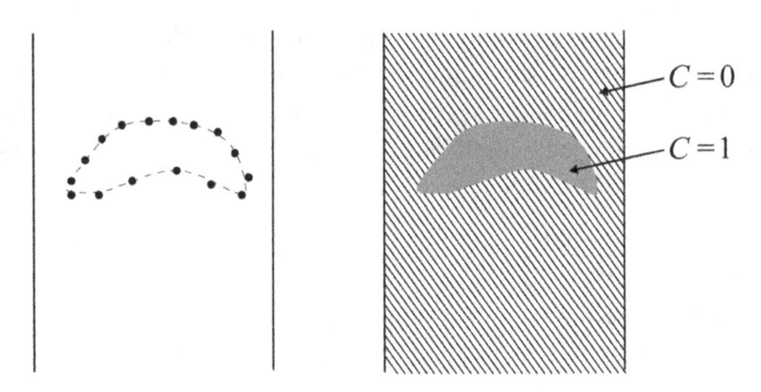

Figure 6.6 A rising bubble from the perspective of the level-set method (left) and the volume-of-fluid method (right).

6.3.6 Models for stratified fluid–fluid flows

One objective in modelling stratified fluid flows is to track the interface, and the Lagrange and Euler–Euler models are not suitable for modelling stratified flows. Instead, front-tracking, level-set or volume-of-fluid methods should be employed, as shown in Figure 6.6. These models may also be employed for DNS of dispersed multiphase flows in cases with a few deformable interfaces, for instance to study the deformation of two colliding droplets. Note that all of these methods assume a no-slip condition at the fluid–fluid interface, and thus have to be resolved to the Kolmogorov length scale. The methods are combined with a single-phase Navier–Stokes type equation,

$$\frac{\partial U_i}{\partial t} + \frac{\partial (U_i U_j)}{\partial x_j} = -\frac{1}{\rho_f}\frac{\partial P}{\partial x_i} + \frac{1}{\rho_f}\frac{\partial \tau_{ji}}{\partial x_j} + g_i, \tag{6.31}$$

where the local fluid properties, e.g. density and viscosity, are given by the presence of the phase,

$$\phi_f(x) = \sum \alpha_k \phi_{fk}. \tag{6.32}$$

If only one phase is present the properties will be for that phase, and a volume-weighted average is used for cells that are shared between two or more phases. The volume fraction is modelled with the continuity equation (6.23). In addition, an extra source term to account for surface-tension effects is also required in these cells.

Although front-tracking, level-set and volume-of-fluid methods are similar, there are some distinct differences, leading to the applicability of each method for different problems. However, only volume-of-fluid methods will be described here.

Volume-of-fluid methods

Volume-of-fluid (VOF) methods use the value of the volume fraction on a grid-cell basis to describe the position of the interface. The advective part of the equation is solved by special advection schemes, such as Lagrangian schemes, geometrical schemes and compressive schemes. These schemes can deal much better with cross-flow situations, and tend to be more mass-conserving than their level-set counterparts. Strictly speaking,

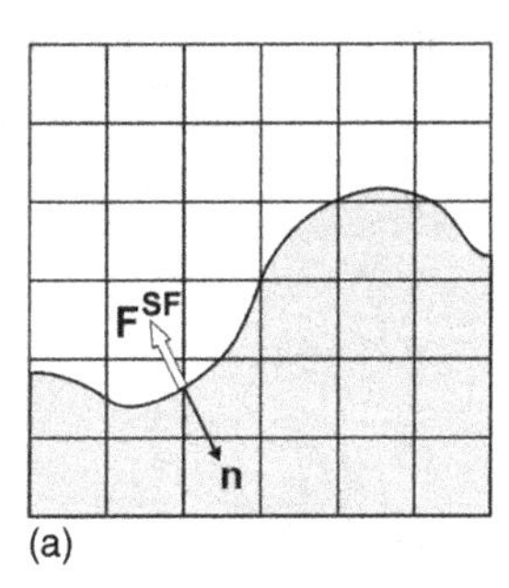

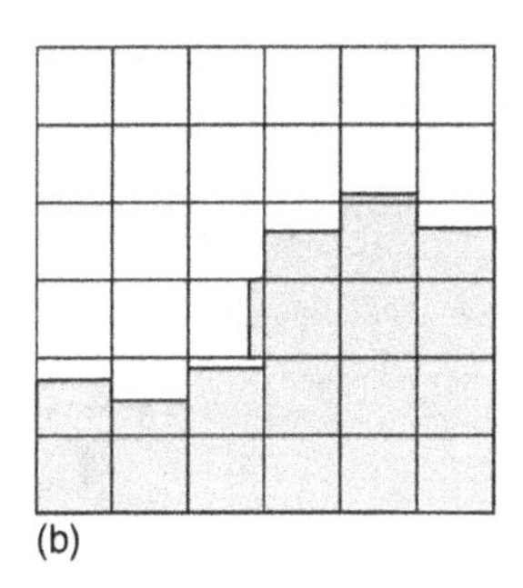

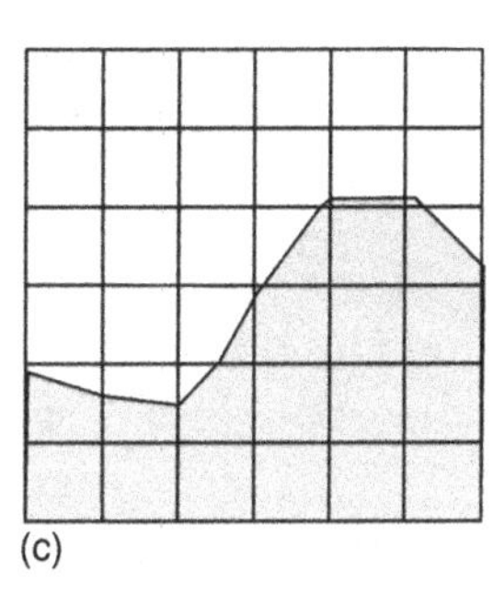

Figure 6.7 Volume-of-fluid (VOF) modelling of a fluid–fluid surface: (a) is the real surface, (b) the volume fraction calculated by the VOF model and (c) the linear reconstruction of the surface.

however, the accuracy is still of first order and a very fine mesh is needed. Typically, about 20 cells/diameter will be needed in order to obtain satisfactory resolution of a spherical bubble or drop.

As soon as an interface between two different fluids ceases to be straight, it has a finite curvature and surface-tension forces may become important. Usually, the surface-tension force is included as an additional momentum source term in a stratified flow model:

$$\frac{\partial(\rho U_i)}{\partial t} + U_i \frac{\partial(\rho U_i)}{\partial x_i} = -\frac{\partial \tau_{ij}}{\partial x_j} + \frac{\partial P}{\partial x_i} + \rho g_i + S_{i,s}. \tag{6.33}$$

It is only in the cells that are shared between the phases that the momentum equation is different from those in the single-phase models. A volume-weighted average of the physical properties according to Eq. (6.32) is used. Numerical instabilities may occur if the properties in the phases are very different, e.g. viscosity ratio $>10^3$.

The direction of the surface-tension force depends upon the interface normals, and its magnitude depends on the interface curvature as shown in Figure 6.7(a). The interface normal, n, in a continuous framework is mostly given as the gradient of the volume fraction,

$$n_i = \frac{\partial \alpha / \partial x_i}{\left| \sum_j \partial \alpha / \partial x_i \right|}. \tag{6.34}$$

To include surface tension between two immiscible, pure fluids in the framework of a front-tracking, VOF or level-set method, usually a continuum surface tension is applied in the form of

$$S_{i,s} = \frac{\sigma \rho \kappa n_i \Gamma}{\frac{1}{2}(\rho_1 + \rho_2)}, \tag{6.35}$$

where σ is the surface-tension coefficient and is usually assumed to be constant, Γ is an interface indicator function and κ is the curvature of the interface. In the most popular continuum surface-tension model, that proposed by Brackbill *et al.* in 1992, the interface indicator, the curvature of the interface and the interface normal are directly related to the volume fraction,

$$\Gamma = \left| \sum_j \frac{\partial \alpha}{\partial x_j} \right|, \tag{6.36}$$

$$\kappa = -\frac{\partial n_i}{\partial x_i}. \tag{6.37}$$

Hence, the curvature is a measure of how fast (magnitude) and in which direction (sign) the normal of an interface changes in space. The surface tension aims at minimizing the interface area. For bubbles or droplets, the minimized interface area is spherical; for a fully stratified flow, the minimized interface area is a straight line. The influence from surface tension can be neglected for capillary numbers $Ca = \mu U/\sigma \gg 1$ or Weber numbers $We = \rho L U^2/\sigma \gg 1$. For large interfaces, such as in stratified flows, but also for interfaces of bubbles or droplets of diameter larger then a few centimetres, the surface-tension force may be negligible. However, for very large curvatures, for example small bubbles with a diameter of a few millimetres, the surface tension may be dominant and in that case ensures that the interface is spherically shaped at all times.

One complication with surface tension not dealt with here is that, in mixtures and with temperature gradients, the surface tension is not constant, and there will appear tangential to the surface a force aiming at minimizing surface energy globally.

Turbulence modelling

In theory the interface should be resolved to the Kolmogorov length scale. This might not be possible in all cases, and standard RANS models must be used with caution. The bulk properties are used, and the damping of turbulence at the interface that is expected when the properties of the phases are very different might not be modelled corectly. Wall functions are not possible since the location of the interface is not given a priori. The k–ω model might be a better choice than the k–ε model, but will require a denser mesh. Large-eddy simulation works better with VOF since the momentum transport across the interface on the subgrid level is much less.

6.3.7 Models for flows in porous media

Flow in a porous medium is driven by the pressure drop over the medium. Examples of such flows occur in fixed-bed reactors, filters and trays in distillation columns. The pressure drop and the resistance of the medium are the most important terms in the governing equations for the fluid phase, i.e. the accelerations are usually small. It is generally not possible to resolve the flow around each object or cavity in a porous medium, so a rough estimate of the pressure drop over the bed is made. The pressure drop in a packed bed originates from viscous resistance, which is proportional to the viscosity and velocity of the fluid phase, and an inertial resistance that is proportional to the density and the velocity squared,

$$\frac{\mathrm{d}P}{\mathrm{d}x_i} = -A\mu U_i - B\left(\frac{1}{2}\rho|U|U_i\right). \tag{6.38}$$

The simplest model for this is called Darcy's law, wherein inertial effects are neglected (i.e. $B = 0$). This is quite frequently used in, for example, simulations of the flow in oil reservoirs. A is then inversely proportional to the area of the pores in the reservoir.

The model of Ergun (dating from 1952) is valid for flow through packed beds for a wide range of velocities and includes both a viscous resistance and an inertial resistance,

$$A = \frac{150(1-\alpha)^2}{d_{\mathrm{p}}^2\alpha^3}, \tag{6.39}$$

$$B = \frac{1.75(1-\alpha)}{d_{\mathrm{p}}\alpha^3}, \tag{6.40}$$

where α is the void fraction and d_{p} is the particle diameter. Note that these values are valid only for flow though beds containing rigid spherical particles. For other types of particles, e.g. fibres and compressible particles, an empirical determination of A and B is required.

6.4 Closure models

Multiphase flow simulations without the requirement of closure models are either extremely computationally expensive, or apply only for very simplified situations. Almost all multiphase flow simulations for engineering applications need closure models to some extent. Most closure models are empirical, meaning that they are determined by experiments and are valid only for the conditions under which the experiments were performed. Some closure models are based upon more theoretical considerations, but most often require assumptions as well. It is important to verify these assumptions and determine whether they are applicable to the simulated conditions. Some of the most popular closure models are discussed here.

6.4.1 Interphase drag

Equation (6.10) is valid for a single particle, i.e. at low volume fraction of the dispersed phase. To model the drag between a fluid and dispersed particles at high volume fraction of the dispersed phase, the drag as formulated in Eq. (6.10) must be adjusted,

$$F_{i,\mathrm{D}} = \frac{1}{2}A_{\mathrm{d}}C_{\mathrm{D}}\rho_{\mathrm{f}}\alpha_{\mathrm{f}}^{-2.65}|U_{i,\mathrm{f}} - U_{i,\mathrm{d}}|\left(U_{i,\mathrm{f}} - U_{i,\mathrm{d}}\right), \tag{6.41}$$

where C_{D} represents the drag coefficient. For spherical particles, this coefficient is a scalar, since it does not depend upon the stream direction of the flow on the particle.

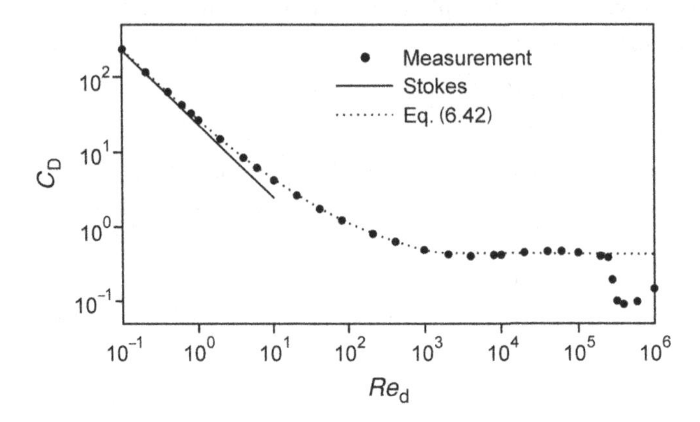

Figure 6.8 The standard drag curve for solid spheres in laminar flows.

Note that the drag is also dependent on particle loading as seen in the α_f dependence in Eq. (6.41). The drag coefficient has different Reynolds-number dependences in the viscous, intermediate and inertial flow regimes, as shown in Figure 6.8, and they are typically divided into three regions,

$$C_D = \begin{cases} \dfrac{24}{Re_d} & \text{if } Re_d < 0.5 \\ \dfrac{24}{Re_d}(1 + 0.15Re_d^{0.687}) & \text{if } 0.5 < Re_d < 1000 \\ 0.44\alpha_f & \text{if } Re_d > 1000, \end{cases} \tag{6.42}$$

where the dispersed-phase Reynolds number, Re_d, is defined as

$$Re_d = \frac{\rho_f D_d |U_f - U_d|}{\mu_f}. \tag{6.43}$$

The drag coefficients in Eq. (6.42) are for particles in laminar continuous phase. The results from experimental studies of drag in turbulent flows have been contradictory, but usually the drag increases with increasing turbulence intensity.

The drag is in general lower for fluid particles, i.e. bubbles and drops due to the circulation of fluid within the particle. For fluid- fluid systems the drag will depend on the viscosity ratio $\kappa = \mu_d/\mu_c$ at low Reynolds numbers:

$$C_D = C_{D0}\left(\frac{\frac{2}{3} + \kappa}{1 + \kappa}\right), \tag{6.44}$$

where C_{D0} is the drag for a solid particle. For gases in pure liquids the drag is a function of *Re* for small bubbles and of the Eötvös number, *Eö*, for large bubbles,

$$C_D = \min\left(\frac{16}{Re}, \frac{8}{3}\frac{E\ddot{o}}{E\ddot{o} + 4}\right), \tag{6.45}$$

where the Eötvös number is defined as

$$E\ddot{o} = \frac{g|\rho_p - \rho_c|D_p^2}{\sigma}. \tag{6.46}$$

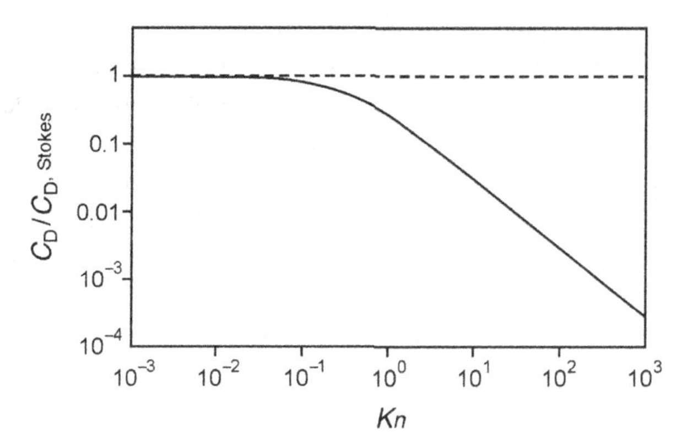

Figure 6.9 The rarefaction effect on drag.

For contaminated systems, e.g. tap water, the surface tension is not constant over the surface, making the bubble more rigid, and the drag coefficient is the same as for a solid particle, i.e. Eq. (6.42).

Much more complex equations are required for non-spherical particles. Large bubbles and drops are deformed and oscillate at high Reynolds number ($Re > 200$), and the drag coefficient cannot be described by a simple function.

In the models above, no slip is assumed for solid particles. In reality most gas molecules adsorb on the surface and will leave the surface in a random direction with the same temperature as the surface. A fraction of the molecules will bounce on the surface and transfer only a part of their momentum to the surface, and they will not be in thermal equilibrium with the surface. However, all molecules will soon collide with other gas-phase molecules, and these molecules will subsequently transfer their added momentum to the surface. The net effect is that the no-slip condition describes the momentum transfer accurately for large particles. However, for small particles or at low pressures the bouncing molecule may collide with other molecules far from the particle, and no secondary collision will occur. This is called the rarefaction effect and will be noticeable when the size of the particle is of the order of the distance between the molecules in the gas, i.e. the mean free path. This is described by the Knudsen number

$$Kn = \frac{\lambda}{d_\mathrm{p}},$$

where λ is the mean free path and d_p the particle diameter. The mean free path is of the order of 75 nm in air at room temperature and 1 atm and rarefaction is important for particles with $d_\mathrm{p} < 1$ μm. Figure 6.9 shows the correction of the standard drag curve due to rarefaction.

6.4.2 Particle interactions

In dense two-fluid flows, particle–particle interactions are important. The interactions can be indirect, whereby the dispersed particles never touch, or direct, whereby the particles actually touch. Examples of indirect interactions occur in bubble columns, between interacting bubbles. However, indirect interactions between two particles may also occur; this is often referred to as the lubrication effect. When the fluid density and viscosity are low, interactions are most often direct.

Models for indirect interactions are required when the behaviour of continuous fluid between the two dispersed particles is not fully resolved. The models depend on how much of the fluid behaviour between the particles is unresolved, which may vary for each simulation. Typically, these models have the form of a long-range potential force, $1/r^{\gamma}$, where r is the distance between the bodies and γ provides a measure for the strength of the interaction. Such models are typically empirical.

Direct interactions typically involve solid particles colliding. There are two characteristic regimes for colliding particles, namely slow granular flow and rapid granular flow. Slow granular flow is characterized by very high volume fractions of solids, low relative velocities between the particles and enduring, multi-body contact. Rapid granular flow is characterized by moderate volume fractions of solids and binary, instantaneous collisions.

Lagrangian particle collisions

When dealing with particle–particle collisions, two types of collision model are widely employed, namely the hard-sphere model and the soft-sphere model. Hard-sphere models describe the dynamics of individual, binary collisions in terms of conservation of momentum and energy. In an ideal collision the momentum is conserved, and the momentum transferred during a collision can be quantified as

$$J = m_1(U_{\mathrm{d},1} - U'_{\mathrm{d},1}) = -m_2(U_{\mathrm{d},2} - U'_{\mathrm{d},2}), \tag{6.47}$$

where m_1 represents the mass of particle 1, $U_{\mathrm{d},1}$ the velocity prior to collision of particle 1 and $U'_{\mathrm{d},1}$ the velocity right after the collision. The post-collision velocities $U'_{\mathrm{d},1}$ and $U'_{\mathrm{d},2}$ can be found by making use of the principle of conservation of energy. More advanced hard-sphere models employ a coefficient of restitution, so part of the kinetic energy of the particles prior to collision is transferred to thermal energy during the collision.

The restitution coefficient describes the damping of velocity at the collision. It is defined as

$$e_{\mathrm{d}} = \frac{U'_{\mathrm{d},2} - U'_{\mathrm{d},1}}{U_{\mathrm{d},1} - U_{\mathrm{d},2}}. \tag{6.48}$$

A restitution coefficient of unity corresponds to ideal collision with no energy losses.

Soft-sphere models aim to estimate the local deformation of the particles during collision. This deformation is due to reversible deformation of the particle. The deformation is related to a linear 'overlap' of the particles during collision. The force resulting from the deformation is typically written in terms of a so-called spring–slider–dashpot model. The spring force is typically the dominant contribution, and is formulated as

$$F = -k\delta_n^{\alpha} n, \tag{6.49}$$

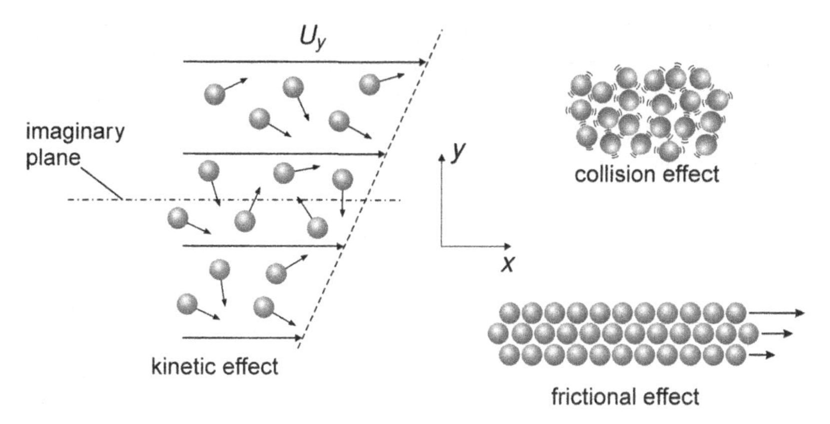

Figure 6.10 Mechanisms for momentum transfer beween particles.

where k is the spring constant, δ represents the overlap and the vector n represents the collision normal, the unit vector attaching the centres of the particles under collision. The exponent α equals 1 for simpler collision models, in which the spring constant cannot be directly related to the physical properties of the particles under collision, whereas a value for α of $3/2$ corresponds to a physically sounder model. The slider and dashpot components of the model represent the irreversibility of the collision, e.g. due to a coefficient of restitution, and the tangential interactions during a collision.

Granular flow models in the continuum framework

Viscosity models for the dispersed phase model the momentum transport between the dispersed particles within the same phase. The interactions with the other phases are described with the interaction coefficients. There are no particle–particle collisions at low loading with one- or two-way coupling, and the granular viscosity will consequently be negligible. The quality of viscosity models at high loading, i.e. with-four way coupling, is very dependent on how ideal the collisions between the particles are. Ideal hard spheres, wood chips or gas bubbles react very differently to collisions, and the quality of the predicted viscosity will vary.

Granular flow models describe the rheology of a suspension of dry granular material. In the continuum framework, granular flow models are based upon simple models for particle interactions, and aim at transforming these interactions towards the continuum scale. Only a brief presentation of the models is given below, with no theoretical derivation of the equations. However, the presentation will give an understanding of what variables are important, e.g. the granular temperature, restitution coefficient and particle pressure.

There are three mechanisms for viscosity, namely particle movement, collision and friction, as shown in Figure 6.10. At low loading with large distances between the particles, most of the momentum transfer occurs by individual particles moving into areas with different average velocities before they collide and transfer momentum. This mechanism is similar to momentum transfer in gases. At higher loading, the particles can move only a short distance before they collide, and most of the momentum transfer occurs by frequent collisions. This mechanism is similar to momentum transfer in liquids. In both these mechanisms, the viscosity depends on the fluctuation motion of the particles, i.e. the granular temperature defined below. At the highest loading the particles are in constant contact and slide over each other. Here the mechanism for momentum transfer is friction, and the viscosity depends on the granular temperature via the particle pressure p_{d}. Granular flow models are usually divided into two families of models, the slow granular flow models and the rapid granular flow models. Slow granular flow is related to momentum transfer by friction and frequent collision of particles in a dense environment, whereas in rapid granular flow the momentum transport occurs by movement and collision of particles.

A popular model in the rapid granular flow regime is the kinetic theory for granular flow [19]. The kinetic theory of granular flow predicts the stresses in moderately dense flows quite accurately, and it has successfully been employed in many applications in this regime. In the rapid granular flow regime, the solids stress arises from particle momentum exchange due to translation and collision. A key parameter in the constitutive closures for the solids phase is the energy associated with the fluctuating motion of the particles, the so-called granular temperature. The granular temperature is defined as the random movement of particles corresponding to how the random movement of molecules in kinetic theory of gases constitutes temperature,

$$\theta = \frac{1}{3}\left(\langle u_1^2\rangle + \langle u_2^2\rangle + \langle u_3^2\rangle\right), \tag{6.50}$$

where u represents the fluctuating velocity of a particle, which has a zero mean by definition. The kinetic theory of granular flows aims to derive a transport equation for the granular temperature. The shearing of the particles causes granular temperature to be produced, and the damping non-ideal component of the collisions, transferring kinetic energy to heat, causes a dissipative effect in granular temperature. The terms in the granular temperature balance arise from collisions and the streaming of particles. In this process, quite a few closures are required, and some of these remain topics for discussion and further research. The balance for granular temperature is very similar to the balance for thermal energy, Eq. (2.28) in Chapter 2. The equation is formulated as

$$\frac{3}{2}\left[\frac{\partial(\rho_{\mathrm{d}}\alpha_{\mathrm{d}}\theta)}{\partial t} + U_{j,\mathrm{d}}\frac{\partial(\rho_{\mathrm{d}}\alpha_{\mathrm{d}}\theta)}{\partial x_j}\right] = \kappa_{\mathrm{d}}\frac{\partial^2\theta}{\partial x_i\,\partial x_j} - p_{\mathrm{d}}\frac{\partial U_{j,\mathrm{d}}}{\partial x_j} + \tau_{kj,\mathrm{d}}\frac{\partial U_{k,\mathrm{d}}}{\partial x_j} - \gamma_{\mathrm{d}}, \tag{6.51}$$

where κ_{d} is the conductivity of granular temperature and the dissipation of fluctuation energy is described by

$$\gamma_{\mathrm{d}} = 3\left(1 - e_{\mathrm{d}}^2\right)\alpha_{\mathrm{d}}^2\rho_{\mathrm{d}}g_0\left(\frac{4}{d_{\mathrm{d}}}\sqrt{\frac{\theta}{\pi}} - \frac{\partial U_{i,\mathrm{d}}}{\partial x_i}\right), \tag{6.52}$$

where e_{d} is the coefficient of restitution for particle collisions. The models presented below are frequently used, but there is no general agreement on a best model. The granular temperature conductivity κ_{d} is a complex function of granular temperature, restitution coefficient etc. that is not given here. The radial distribution function at particle contact is

$$g_0 = \left[1 - \left(\frac{\alpha_\mathrm{d}}{\alpha_{\mathrm{d,max}}}\right)^{1/3}\right]^{-1}, \tag{6.53}$$

where the maximum volume fraction of the dispersed phase $\alpha_{\mathrm{d,max}}$ is used to calculate a dimensionless particle–particle distance. This function (g_0) also certifies that the maximum particle loading is never surpassed by increasing the particle pressure to infinity when $\alpha_\mathrm{d} \to \alpha_{\mathrm{d,max}}$. The total particle pressure due to the streaming and collisions of particles is given as

$$p_\mathrm{d} = \alpha_\mathrm{d}\rho_\mathrm{d}\theta + 2\rho_\mathrm{d}(1 + e_\mathrm{d})\alpha_\mathrm{d}^2 g_0\theta. \tag{6.54}$$

The kinetic and collision parts of the granular viscosity can be formulated as

$$\mu_{\mathrm{d,kin}} = \frac{\alpha_\mathrm{d} d_\mathrm{d} \rho_\mathrm{d}\sqrt{\theta\pi}}{6(1 - e_\mathrm{d})}\left[1 + \frac{2}{5}(1 + e_\mathrm{d})(3e_\mathrm{d} - 1)\alpha_\mathrm{d} g_0\right], \tag{6.55}$$

$$\mu_{\mathrm{d,coll}} = \frac{4}{5}\alpha_\mathrm{d}\rho_\mathrm{d} d_\mathrm{d} g_0(1 + e_\mathrm{d})\left(\frac{\theta}{\pi}\right)^{1/2}. \tag{6.56}$$

Slow granular flow models are based upon theories of soil mechanics, including a global inertia effect in the governing equations of the solids. The volume fractions of slow granular flows are typically very large, just below the maximum packing of the particles. Therefore, gradients of the volume fraction are typically neglected. The random kinetic energy of individual particles, the granular temperature, is very low and neglected as well. The granular flow is dominated by frictional stress, and the classical slow granular flow models based upon Mohr–Coulomb models have successfully been employed to predict the particle flow in, for instance, hoppers and chutes. Such models describe how normal pressure and viscosity depend upon the strain-rate tensor; for instance, for the frictional viscosity

$$\mu_{\mathrm{d,fr}} = \frac{p_\mathrm{d}\sin\phi}{2\sqrt{I_{2\mathrm{D}}}}, \tag{6.57}$$

where $I_{2\mathrm{D}}$ is related to the principal strain-rate tensor. Some parameters must be set in the model, i.e. the coefficient of restitution for particle collisions e_d and the angle of internal friction ϕ. The angle of internal friction ϕ is the maximum angle of a stable slope determined by friction, cohesion and the shapes of the particles.

Although rapid granular flow occurs less often in nature than slow granular flow, there are much sounder models for this flow regime. The basic assumption for such models is that particle interactions comprise binary, instantaneous and nearly elastic collisions. The particle rheology is directly expressed in the granular temperature. Since the number of particles is strongly dependent upon the solids volume fraction, the particle stresses have a strong dependence upon the solids volume fraction. When the solids volume fraction reaches the maximum packing, the number of collisions is infinite, and so are the predicted particulate stresses. Rapid granular flow models are not very accurate in this very dense regime, since the inter-particle interaction is no longer instantaneous,

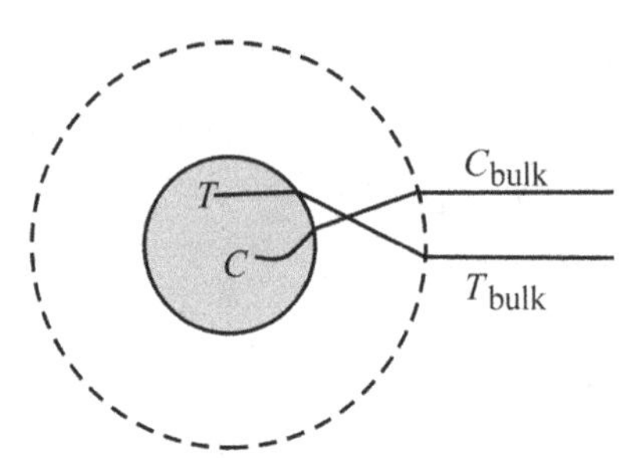

Figure 6.11 Mass and heat transfer between a particle and the bulk fluid.

and a transition to slow granular flow models should be made. For example, adding the stresses arising from the rapid granular flow regime to the slow granular flow regime,

$$\mu_{\text{d}} = \mu_{\text{d,kin}} + \mu_{\text{d,col}} + \mu_{\text{d,fr}}$$

gives a model that can handle the whole range of particle loading reasonably well.

6.4.3 Heat and mass transfer

Heat and/or mass transfer often plays an important role in multiphase flows. Examples are evaporating liquid droplets, reactions on catalysts, burning of coal particles and the creation of bubbles by means of cavitation. Models describing mass and heat transfer are typically quite crude and are built upon empirical or simplified relations of the Sherwood or Nusselt dimensionless parameters. The particles may be inert particles that can be heated or cooled, droplets that can evaporate or condense, fuel particles that may burn or catalyst particles that catalyse reactions.

When the net flow from or to the particle is small, mass transfer between a particle and the bulk fluid can be estimated using the film model. The mass flux [mol m^{-2} s^{-1}] is then expressed as

$$N_n = k_{\text{c},n} \left(C_{n,\text{s}} - C_{n,\text{bulk}} \right), \tag{6.58}$$

where $C_{n,\text{s}}$ is the concentration of species n on the surface and $C_{n,\text{bulk}}$ is the concentration in the fluid bulk surrounding the particle (Figure 6.11). The mass-transfer coefficient $k_{\text{c},i}$ is often described by a Sherwood number, Sh,

$$Sh = \frac{k_{\text{c},i} d_{\text{p}}}{D_i} = 2 + 0.6 Re_{\text{d}}^{1/2} Sc^{1/3}. \tag{6.59}$$

According to this correlation the mass transfer between particles and a fluid depends on the molecular diffusivity, the size of the particles, the viscosity and the relative velocity between the fluid and the particles. One advantage using CFD is that the local Re_{d} and local Sh are obtained. However, evaporation, condensation and chemical reactions cause a net flow to or from the particle that may decrease or increase mass and heat transfer, and Eq. (6.58) must be adjusted accordingly. The drag is also affected by the net flow to and from the particles.

Heat transfer is described in a similar way using the Nusselt number, Nu,

$$Nu = \frac{hd_{\mathrm{p}}}{\lambda} = 2 + 0.6Re_{\mathrm{d}}^{1/2}Pr^{1/3}, \tag{6.60}$$

and the heat balance of a particle is written

$$m_{\mathrm{p}}c_{\mathrm{p}}\frac{\mathrm{d}T_{\mathrm{p}}}{\mathrm{d}t} = hA_{\mathrm{p}}\left(T_{\mathrm{bulk}} - T_{\mathrm{p}}\right) + \frac{\mathrm{d}m_{\mathrm{p}}}{\mathrm{d}t}h_{\mathrm{fg}} + R\Delta H + A_{\mathrm{p}}\varepsilon_{\mathrm{p}}\sigma(T_{\mathrm{surr}}^4 - T_{\mathrm{p}}^4), \tag{6.61}$$

where m_{p} is the mass of the particle, h_{fg} the enthalpy of evaporation, R the reaction rate and ΔH the reaction enthalpy. In this balance we have included the accumulation of heat in the particle, the decrease in particle weight due to evaporation with corresponding heat loss, the heat due to a reaction in the particle and radiation from the particle to the surrounding.

In boiling, condensation and cavitation phase changes occur. This requires an additional term in the equation for conservation of mass,

$$\frac{\partial(\alpha_k\rho_k)}{\partial t} + \frac{\partial(\alpha_k\rho_k u_{i,k})}{\partial x_i} = \dot{m}_k, \tag{6.62}$$

where $\dot{m}_k$ represents the mass transfer from or to phase k per unit volume and unit time.

6.5 Boundaries and boundary conditions

When performing simulations with multiphase flows, boundary conditions should be given for each phase at each boundary. The boundary conditions for the continuous phase resemble the type of boundary condition for single-phase flows. At an inlet, a velocity profile or mass flow rate may be specified. Additionally, a volume fraction is often specified. If a turbulence model is employed for the continuous phase, the turbulence properties of the flow entering the domain should also be specified. At the outlet, typically a pressure or reference pressure is specified. If there is no backflow entering the outlet, the setting of most flow properties is not very important. If there is a flow entering the domain through the outlet, some of the outlet boundary conditions may be specified in a manner similar to what is done for an inlet.

At walls usually a 'no-slip' boundary condition is specified for the continuous phase, with additional wall functions when the flow is turbulent. Particle–wall collisions are of importance in most confined flows. Because particle–wall collisions are non-ideal, a particle–wall impact is associated with a deceleration of the particle, which is re-accelerated on re-entering the flow. Hence, a particle–wall impact indirectly extracts momentum from the continuous phase, causing a pressure loss. This effect is enhanced in the case of 'soft' walls.

6.5.1 Lagrangian dispersed phase

In Lagrangian particle simulations, particle–wall collisions are treated by using models similar to those for particle–particle collisions. If a hard-sphere model is adopted, the

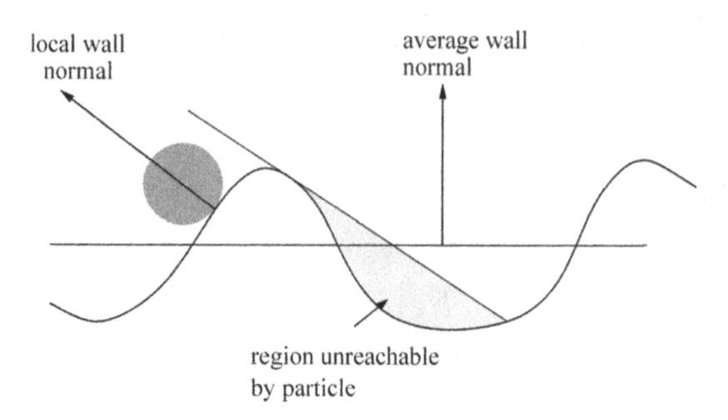

Figure 6.12 A particle colliding with a rough wall.

particle–wall collision may dissipate kinetic energy as heat that can be estimated from the coefficient of restitution. The restitution coefficient is defined as the ratio of the velocity after collision and that before collision. If a soft-sphere model is adopted, the effect of the wall may be described using a spring–slider–dashpot model based upon heuristics or upon a number of physical parameters. Note that, in the application both of hard-sphere and of soft-sphere models, the properties of the wall are not necessarily the same as the properties of the particles. Most often, a wall can be treated as a particle with infinite mass and radius.

Rough walls

Experiments and simulations have shown that wall roughness may have a considerable effect on the particle–wall collision process, and thus on the complete flow characteristics. This has two causes. First, the normal of the actual wall might not be constant across the wall. This means that the impact normal is a function of the local wall roughness. This can be modelled by including a 'random' component on top of the average wall normal. Secondly, the so-called shadow effect results in a shift of the effective roughness distribution for small impact angles. Both effects are shown in Figure 6.12: (1) the local wall normal changes along the wall, affecting the collision normal; and (2) particles with small impact angles can no longer collide with parts of the wall, making certain collision normals more likely than others (this was called the shadow effect by Sommerfeld in 1996).

6.5.2 Eulerian dispersed phase

Boundary conditions for walls in Eulerian models are less obvious than in the Lagrangian framework. In most simulations, particles are assumed to slip freely along the walls. This assumption originates from experimental observations, in which most particles seem to move rather freely along walls. More realistically, a slip condition can be specified at the walls,

$$\frac{\partial U_{i,\mathrm{d}}}{\partial n_i} = f, \tag{6.63}$$

where f specifies the amount of slip. If $f = 0$, the free-slip conditions are obtained. It is generally observed that convergence becomes difficult for higher values of f.

Boundary conditions for the granular temperature at the wall may have a significant impact on the flow. Various models that mimic the wall functions for single-phase turbulent flows are available. These have a production term, accounting for the dispersion effect of the particles, and a dissipation term, accounting for the non-ideal effect of the collision. In most simulations, specifying a zero gradient of granular temperature at walls gives satisfactory results.

6.6 Summary

In this chapter the Euler–Lagrange, Euler–Euler, mixture, volume-of-fluid and packed-bed models are presented. In many cases, it is not evident which model to choose. Each of the models is based on a number of assumptions and provides a specific level of detail. Some models describe individual particles or interface segments, whereas other models are expressed in terms of averaged properties. When selecting a model, numerical stability and computational cost must also be considered.

The *Euler–Euler* model is obtained by averaging the flow properties of a multiphase flow of dispersed particles smaller than the computational cells. The model solves the momentum and continuity equations for each phase. It is the most general model and should in principle work for all kinds of multiphase flows. However, it is usually less stable than the other models, and it requires closure models to describe all interactions between the phases, so empirical models for the specific system are often required. The *mixture* model is a simplification of the Euler–Euler model that is more stable but requires a strong coupling between the phases, i.e. low Stokes number. It shares with the Euler–Euler model the problem that empirical closures are required for any interaction between the phases.

In the *Euler–Lagrange* model the transport equations for each particle or a bundle of particles that behave identically are solved. It is most accurate and works very well for systems with a limited number of particles, often in cases in which particle–particle collisions can be neglected, i.e. when the particle–particle distance is larger than ten particle diameters. In principle, inter-particle collisions can be taken into account, but this is usually very expensive. The upper limit for the number of particles or bundles of particles is a few hundred thousand when the simulations are done on a PC. The model tracks the centre of gravity of the bundle, and a limit for the bundle size is that most of the particles in the bundle should be within one computational cell if one is to describe the interaction with the continuous phase accurately. The Euler–Lagrange model can easily take into account particle-size or density distributions. For a large size or density distribution the Euler–Euler model requires several phases to describe the system and relations to describe their interactions. In these cases, the Euler–Lagrange model will be more stable and will predict the flow better.

The *volume-of-fluid* (VOF) model is intended for gas–liquid and liquid–liquid segregated flows. The model resolves the interface between the phases and requires that the

Table 6.3 Suggested multiphase models for some common processes

System	Criterion	Example	Model
Gas/liquid–solid	Few particles or bundles of particles. Particle–particle distance $<10d_p$	Sedimentation	Euler–Lagrange
	Many particles $St < 1$	Stirred tank	Mixture
	Many particles $St > 1$	Fluidized bed Cyclone sedimentation	Euler–Euler
Bubbly flow	Few bubbles, bubble–bubble distance $>10d_p$		Euler–Lagrange
	Large volume fraction of gas	Bubble column	Mixture
Segregated gas–liquid	Large fluid particles	Slugs, waves	Volume of fluid
Sprays	Few drops or bundles of drops	Spray	Euler–Lagrange
Liquid–liquid	Stokes number $St < 1$	Extraction	Mixture
	Stokes number $St > 1$		Euler–Euler
	Large fluid particles		Volume of fluid
Gas–liquid–solid	Stokes number $St > 1$	Slurry reactor	Mixture
	Stokes number $St < 1$	Flotation	Euler–Euler

fluid particles, e.g. droplets or bubbles, are much larger than the computational cells. The phases are modelled using a single velocity field, and the computational cells that are shared between the phases have to be averaged. The model works well in flows with not too high a viscosity or density ratio between the phases. Since the method requires a DNS resolution of the flow at the interface, the damping of turbulence at the interface in turbulent flows is not well modelled and the VOF model may overestimate the shear stress at the interface. The model requires a large number of mesh cells, and Cartesian mesh cells make the method much more accurate.

The *porous-bed* model predicts flow in channels and cavities much smaller than the mesh. The model requires that the volume fraction of the continuous phase is given a priori and that the pressure drop is given by an algebraic model.

6.6.1 Guidelines for selecting a multiphase model

In selecting the best multiphase model, the first step is to identify the porous domains and attribute appropriate flow models to each area. The second step is to characterize the flow as segregated or dispersed. Other parameters that are important in selecting the best model are the particle loading and the Stokes number. The particle loading will give an estimate for the number of particles and the probability of particle–particle collisions. The Stokes number predicts how independently the dispersed phase behaves relative to the continuous phase. The scheme in Figure 6.13 gives a crude view of the choices. However, the choices are not clear-cut insofar as there might be other reasons for selection of models (Table 6.3). Very often it is the stability of the solution and available data that decide the selection.

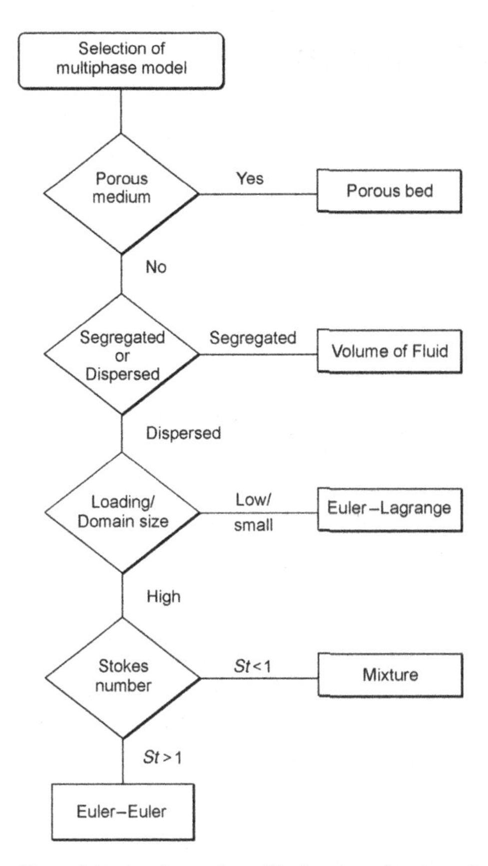

Figure 6.13 A schematic guide for the selection of multiphase models.

Questions

(1) Describe a few key parameters that are useful for characterizing multiphase flows.
(2) What is the physical interpretation of the Stokes number?
(3) Several forces act on a single particle in a fluid flow. Discuss the most important of these.
(4) Discuss what is meant by phase coupling.
(5) Explain what models it is appropriate to use for different types of dispersed phase flows and stratified flow.
(6) What closure models are required for multiphase flow simulations?
(7) Explain the concept of viscosity modelling for granular flows.

7 Best practice guidelines

Computational fluid dynamics does not provide an exact solution to all problems, but is in many cases a reliable tool that can provide useful results when it is employed by an experienced user. An inexperienced user, on the other hand, may obtain very nice graphs that are very far from being a prediction of the stated problem. Some of the problems arise from the many default settings in commercial CFD codes, since the user may obtain results without knowing what the code is doing by accepting settings that are not appropriate for the specific problem. The user must make an active decision regarding each setting due to the fact that many problems can arise from a user failing to understand what the proper settings should be. This chapter provides some guidelines that can help a new user to avoid the most common mistakes. Many more recomendations selected by experienced CFD users can be found in the 'Best Practice Guidelines' for single-phase flows [20] and for dispersed multiphase flows [21] by the European Research Community on Flow Turbulence and Combustion (ERCOFTAC).

A CFD simulation contains both errors and uncertainties. An *error* is defined as a recognizable deficiency that is not due to a lack of knowledge, whereas an *uncertainty* is a potential deficiency that is due to a lack of knowledge.

Poor simulation results may arise for many different reasons; the user may make mistakes in formulating the problem to solve or in formulating the geometry and meshing. Numerical errors may occur due to lack of convergence and a poor choice of discretization methods. The models for turbulence, reactions and multiphase flow are not exact descriptions of the real world, and poor results may be introduced by not selecting the best model for the specific case. In the end, the robustness and reliability of the simulations should be analysed by sensitivity analysis, verification, validation and calibration.

In general, it is good practice to study the obtained simulation results critically. CFD is not (yet) at the stage at which it can be treated as a black box. Simulation results should be verified with experimental findings, fluid-mechanics theory and, sometimes, instinct. It may also help to understand the limitations of the CFD model if a simulation is performed for a case regarding which the results are known, prior to exploring new ground with the model. Some common errors and recommendations for best practice are given in this chapter.

7.1 Application uncertainty

Many problems in CFD simulations arise from inaccurately stated problems. The actual problem that is being solved is not always straightforward. It is seldom possible to select the right settings from the beginning. The properties of the flow are not all known in advance. The flow can be laminar or turbulent, or laminar in some regions and turbulent in other regions. The local hold-up in multiphase flow and the local Dahmköhler number in reacting systems can affect the choice of model. The flow within the system may also affect the flow at the inlet and the outlet. These are just a few of the possibilities.

7.1.1 Geometry and grid design

A number of general guidelines can be given.

- Make sure that the CAD geometry is complete for the flow simulation. The CAD drawing should be kept as simple as possible, but it must not be simpler than that. All details smaller than the computational cells can in most cases be removed, but small details on the surface can sometimes be important for the flow, e.g. a welded joint may induce flow separation.
- Symmetric boundary conditions restrict the solution to a symmetric solution, and no transport is allowed across the symmetry plane, e.g. bubble columns and fluidized beds are poorly described by 2D simulations.
- When the inflow and outflow are not known exactly, they should be put far from the region of interest.
- With constant-pressure outflow, specify the direction of the outflow (e.g. normal to the plane) to minimize the pressure difference across the surface.
- Use pressure outflow for multiple outflow boundaries.
- Avoid having non-orthogonal cells close to the boundaries. The angle between the grid lines and the boundary should be close to 90°. Use body-fitted grids when the grid cannot be aligned with the surface.
- Avoid the use of highly skewed cells. The angles should be kept between 40° and 140°. The maximum skewness should be <0.95 and the average below 0.33. The aspect ratio should be below 5, but may be up to 10 inside the boundary layer.
- The squish index describes how the cell faces are oriented, where 0 is a perfect cell and the upper limit is 1. The recommendation is that the maximum squish index should be below 0.99.
- Perform a grid check to avoid problems due to incorrect connectivity. This can often be done in the mesh-generator software.

7.2 Numerical uncertainty

7.2.1 Convergence

Poor convergence is the most common numerical reason for poor results. Residuals are defined differently in various programs, and monitoring the residual is in most cases not sufficient to verify that a converged solution has been reached. Convergence problems for steady-state simulations may also indicate that a steady solution does not exist. The residual should always be combined with other measures of convergence. Keep in mind that a converged solution is not always a correct one.

- Use different convergence criteria for different variables. Concentration may need a lower residual than velocity.
- Monitor integral quantities of solution-sensitive variables.
- Make global balances for mass, momentum and energy.
- Monitor the solution at specific important points.
- Test for steady state by switching to a transient solver.
- Plot the residual to evaluate whether the solution is poor in some regions of the computational domain.

7.2.2 Enhancing convergence

There are many different methods by which to enhance convergence.

- Use more robust numerical schemes, e.g. initially employ first-order upwind, changing to a higher order for the final iterations.
- Reduce the under-relaxation or CFL number initially.
- Examine the local residual. The convergence problem might be localized to one small region. Use grid adaptation to refine or coarsen the mesh in areas where it is needed. Using a fine grid throughout may diminish the convergence rate.
- Solve steady-state problems transiently.
- Try different initial guesses, e.g. obtain an initial guess from a short transient simulation or obtain an initial guess from a steady-state simulation.
- Solve for only a few variables at a time, e.g. solve for the flow field first and keep the velocities constant whilst solving concentration and chemical reactions. Finally solve for all variables.
- Use the coupled solver for high-speed compressible flows, highly coupled flows with strong body forces or flows being solved on very fine meshes. Keep in mind that the coupled solver requires 1.5–2 times more memory than needed by a segregated solver.

7.2.3 Numerical errors

The numerical error is the difference between the exact solution and the numerical solution.

- Avoid the use of first-order schemes. First order can be used initially when you have convergence problems, but will always lead to problems with numerical diffusion.
- Estimate the discretization error by showing that the solution is independent of the mesh density.
- Use node-based gradients with unstructured tetrahedral meshes.
- If possible, compare the solutions with different orders of accuracy.

- Perform refinement or coarsening of the computational mesh. Ideally, the solution should become less dependent on the mesh as it is refined. If the results reveal a strong dependence on the computational mesh spacing, further refinement is required.

7.2.4 Temporal discretization

The choice of time step depends on the flow features being studied. Large-eddy simulation requires very short time steps to resolve the turbulent eddies. Some of the turbulent structures can also be resolved with RANS models, i.e. unsteady RANS or URANS, but usually the time step is much larger than the turbulence timescales and only the transient behaviour of the mean flow is resolved.

- Start with a short time step, i.e. a low CLF number that can increase with time.
- The time step should be selected so that fewer than 20 iterations are needed in each time step with an implicit solver.
- Make sure that the solution has converged for each iteration for implicit schemes.
- For temporal accuracy, a second-order-accurate discretization scheme is recommended, such as Crank–Nicolson or second-order backward Euler.

7.3 Turbulence modelling

Only DNS is an exact formulation of the Navier–Stokes equations. All other turbulence models contain model approximations. Table 4.4 contains a comparison of turbulence models.

- Test different turbulence models. The cost of running a second model starting from a converged solution is usually not high. Different values of the model constants that are recommended for specific kinds of flows should also be tested.
- Transition between turbulent and laminar flow is very difficult to simulate for all models. Simulations involving transition must be verified with experimental data.
- Be aware of the limitations of the specific model, e.g. the k–ε model will suppress swirl in a flow and the absence of swirl in a k–ε model does not certify that swirl would not appear with another turbulence model.
- Many physical phenomena are not captured by turbulence models and require understanding from the user. Examples include, but are not limited to, transition, separation, unsteady boundary layers and low-Reynolds-number turbulence.

7.3.1 Boundary conditions

- The exact inlet conditions for the turbulence properties are usually not known exactly. The value of k/ε times the average velocity gives an order-of-magnitude estimation of how far into the system the settings at the inlet will survive.
- Exact inlet conditions for LES are not possible. Modern CFD programs can generate simple turbulent eddies at the inlet, but the inlet should be far from the area of

interest in order to allow a proper statistical distribution of eddies to develop. Periodic boundary conditions should be used when possible, but may introduce unphysical correlations. As always, it is important to verify the obtained solution critically.

- For RANS models, the lower level of y^+ should typically be between 20 and 30 at the walls. Some commercial CFD programs can be made to accept lower y^+ levels by adjusting the appropriate model. (Check the manual for actual values.) The upper limit of y^+ is usually in the range 80–100.
- LES models require additional treatment at no-slip walls.
- For low-*Re* turbulence models the first grid should be at $y^+ < 4$, preferably at $y^+ \approx 1$ with 5–10 mesh points below $y^+ = 20$.
- Standard wall functions are not recommended for flow with a negative pressure gradient, e.g. with flow separation at the wall.

7.4 Reactions

- Check the Dahmköhler number. Slow chemical reactions with $Da \ll 1$ are straightforward. Very fast, mixing-controlled, isothermal chemical reactions with $Da \gg 1$ can be modelled rather accurately. Most other conditions with $Da \approx 1$ and reactions that lead to heat formation and changes in density will give uncertain results.
- Very low residuals as a convergence criterion for concentration are often required, and monitoring integral quantities for mass balances or steady local concentrations usually provides more reliable indicators than low residuals.

7.5 Multiphase modelling

Generally speaking, multiphase flows are more challenging than single-phase flows, and errors in multiphase flow simulations are typically larger than those in single-phase simulations. These errors may have a number of different origins.

- **Not knowing the most important physical mechanisms.** Because of the wide variety of multiphase flow types, there is not one 'generic' model for multiphase flows. Before attempting to model a multiphase flow system, it is very important to understand the physical mechanisms occurring in the flow system. This includes understanding the most important forces and mechanisms in the flow and the properties of the fluid(s) and/or solids, as well as a good estimate of the length scales and timescales of the physical processes. Only with this knowledge can appropriate models be selected and their shortcomings in a simulation estimated.
- **Closure models.** Most errors in multiphase flow simulations arise from shortcomings of the closure models employed. Most closure models are empirically determined, which makes them applicable, strictly speaking, only under conditions similar to those for the data they are built from. Analytical closure models are developed for 'ideal' conditions, which are hardly ever met in reality. Using closure models for

conditions or regimes different from those to which they are applicable may be asking for trouble.

- **Timescale and length-scale separation.** In the derivation of turbulence models, a separation between the timescales and length scales of the 'large' eddies carrying the energy and the scale(s) at which energy is dissipated is assumed. Similar assumptions are required in order to derive governing and closure models for multiphase flows. For instance, many closure models require the particles to be smaller than the large-scale flow structures in a flow.
- **Choice of model and governing equations.** It is important to understand the implications and assumptions of each multiphase model, and under what conditions it may be employed. For instance, in the derivation of the Euler–Euler or two-fluid model, the pressure gradient over large interfaces, which is important in separated-flow situations, is neglected. This means that such a model will not capture the dynamics of free surfaces very accurately.
- **Numerical errors.** Many numerical errors potentially arising in single-phase calculations may also arise in multiphase flow calculations. Therefore, it is important to check the best-practice guidelines for single-phase flow computations. However, there may be additional problems. Many of the ideas employed in solving multiphase flows arise from single-phase flows, leading to slow convergence, or, worse, an erroneous result.

To minimize the potential problems occurring when performing multiphase flow simulations, the following checklist may be employed,

1. If possible, start the simulation with a single-phase flow situation resembling the system. This simulation can be optimized in terms of grid size, time step, etc. by using the best-practice guidelines for single-phase flow.
2. Determine the regime of the multiphase flow in terms of dimensionless parameters (*Re*, *We*, *St*, . . .). This enables the choice of suitable closure models and may give insight into the expected flow situation.
3. Make an estimate of the forces acting on bubbles, particles or droplets and under which conditions these forces will occur.
4. Make a suitable selection of the turbulence model and decide which terms (and coupling to the dispersed phase) are important.
5. If possible, start with a geometry, flow properties and dispersed-phase properties similar to those of a system of which you know the behaviour or for which experimental data are available. This creates confidence in the models employed.
6. If there is a large size distribution of the dispersed phase, a multi-fluid approach might be required for the dispersed phase. This allows the use of a range of size classes, which can be monitored separately. Size distributions can have a big effect on the flow.
7. First-order-accurate models, such as the VOF model and the surface-tension models, require a very fine mesh – in these cases a relatively mesh-independent solution is very important.

8. Make sure that iterations are well converged. Many popular commercial CFD solvers will start with a new time step when a specified maximum number of iterations is reached, regardless of convergence criteria. This may be detrimental in terms of the quality of results obtained.

7.6 Sensitivity analysis

There are many choices that are uncertain, and sensitivity analysis is one way to examine the effect of the specific choice. It is good practice to start from a converged solution and change all uncertain settings in the model.

7.7 Verification, validation and calibration

Verification is a procedure to ensure that the program solves the equations correctly. Validation is done to test how accurately the model represents reality, and calibration is often used to adjust the simulation to known experimental data in order to study parameter sensitivity ('what if') in the design process.

Appendix

Einstein or tensor notation

In mathematics, especially in applications of linear algebra to fluid mechanics, the **Einstein notation** or **Einstein summation convention** is a notational convention that is useful when dealing with coordinate equations or formulas.

According to this convention, when an index variable appears twice in a single term, it implies that we are summing over all of its possible values. In typical applications, these are 1, 2 and 3 (for calculations in Euclidean space).

Definitions

In the traditional usage, one has in mind a vector space V with finite dimension n, and a specific basis of V. We can write the basis vectors as $\mathbf{e}_1, \mathbf{e}_2, \ldots, \mathbf{e}_n$. Then, if $\mathbf{v}$ is a vector in V, it has coordinates $v_1, \ldots, v_n$ relative to this basis.

The basic rule is

$$(\mathbf{v})_i = v_i \mathbf{e}_i .$$

In this expression, it is assumed that the term on the right-hand side is to be summed as i goes from 1 to n, because the index i appears twice. The index i is known as a *dummy index* since the result is not dependent on it; thus we could also write, for example,

$$(\mathbf{v})_j = v_j \mathbf{e}_j .$$

An index that is not summed over is a *free index*, and should be found in each term of the equation or formula.

If $\boldsymbol{H}$ is a matrix and $\mathbf{v}$ is a column vector, then $\boldsymbol{H}\mathbf{v}$ is another column vector. To define $\mathbf{w} = \boldsymbol{H}\mathbf{v}$, we can write

$$w_i = H_{ij} v_j .$$

The dot product of two vectors $\mathbf{u}$ and $\mathbf{v}$ can be written

$$\mathbf{u} \cdot \mathbf{v} = u_i v_i .$$

There are two useful symbols that simplify multiplication rules, the

Kronecker delta,

$$\delta_{ij} = (\mathbf{I})_{ij} = \begin{cases} 1 & \text{if } i = j \\ 0 & \text{if } i \neq j, \end{cases}$$

and the Levi-Civita symbol e (or ε),

$$e_{ijk} = \begin{cases} 1 & \text{if } (ijk) \text{ is a positive permutation of } (1, 2, 3) \\ -1 & \text{if } (ijk) \text{ is a negative permutation of } (1, 2, 3) \\ 0 & \text{if } (ijk) \text{ is a not a permutation of } (1, 2, 3) \text{ at all;} \end{cases}$$

(1, 2, 3), (3, 1, 2) and (2, 3, 1) are positive, (3, 2, 1), (1, 3, 2) and (2, 1, 3) are negative and (1, 2, 2) etc. are not permutations of (1, 2, 3).

If $n = 3$, we can write the cross product, using the Levi-Civita symbol. Specifically, if $\mathbf{w}$ is $\mathbf{u} \times \mathbf{v}$, then

$$w_i = e_{ijk} u_j v_k.$$

Operators

For general operations on scalars, vectors and matrices in fluid mechanics, ϕ is any scalar having the rank 0, $\mathbf{u}$ is a velocity vector

$$\mathbf{u} = \begin{bmatrix} u_1 \\ u_2 \\ u_3 \end{bmatrix}$$

having the rank 1 and $\boldsymbol{\tau}$ is a (3×3) tensor

$$\boldsymbol{\tau} = \begin{bmatrix} \tau_{11} & \tau_{12} & \tau_{13} \\ \tau_{21} & \tau_{22} & \tau_{23} \\ \tau_{31} & \tau_{32} & \tau_{33} \end{bmatrix}$$

having the rank 2.

Various operators will change the rank of the tensor:

- gradients will increase the rank by 1,
- $\times$ product decreases the rank by 1,
- $\bullet$ product decreases the rank by 2,
- : product decreases the rank by 4.

The gradient of a scalar is

$$\nabla\phi = \begin{bmatrix} \dfrac{\partial\phi}{\partial x_1} \\ \dfrac{\partial\phi}{\partial x_2} \\ \dfrac{\partial\phi}{\partial x_3} \end{bmatrix}$$

or

$$(\nabla\phi)_i = \frac{\partial}{\partial x_i}\phi.$$

The rank is $0 + 1 = 1$.

The Laplacian is

$$\nabla\cdot\nabla\phi = \nabla^2\phi = \Delta\phi = \frac{\partial^2\phi}{\partial x_1^2} + \frac{\partial^2\phi}{\partial x_2^2} + \frac{\partial^2\phi}{\partial x_3^2} = \left(\frac{\partial^2}{\partial x_1^2} + \frac{\partial^2}{\partial x_2^2} + \frac{\partial^2}{\partial x_3^2}\right)\phi$$

or

$$\Delta\phi = \frac{\partial^2\phi}{\partial x_i\,\partial x_i}.$$

The rank is $0 + 1 - 1 = 0$.

For $\mathbf{u}$

$$\nabla^2\mathbf{u} = \begin{bmatrix} \dfrac{\partial^2 u_1}{\partial x_1^2} + \dfrac{\partial^2 u_1}{\partial x_2^2} + \dfrac{\partial^2 u_1}{\partial x_1^2} \\ \dfrac{\partial^2 u_2}{\partial x_1^2} + \dfrac{\partial^2 u_2}{\partial x_2^2} + \dfrac{\partial^2 u_2}{\partial x_3^2} \\ \dfrac{\partial^2 u_3}{\partial x_1^2} + \dfrac{\partial^2 u_3}{\partial x_2^2} + \dfrac{\partial^2 u_3}{\partial x_3^2} \end{bmatrix}$$

or

$$(\Delta\mathbf{u})_i = \frac{\partial^2 u_i}{\partial x_j\,\partial x_j}.$$

The rank is $1 + 1 - 1 = 1$.

For the dot product, the divergence,

$$\nabla\cdot\mathbf{u} = \frac{\partial u_1}{\partial x_1} + \frac{\partial u_2}{\partial x_2} + \frac{\partial u_3}{\partial x_3}$$

or

$$\nabla \cdot \mathbf{u} = \frac{\partial}{\partial x_i} u_i;$$

$$\mathbf{u} \cdot \nabla \mathbf{u} = \begin{bmatrix} u_1 \frac{\partial u_1}{\partial x_1} + u_2 \frac{\partial u_1}{\partial x_2} + u_3 \frac{\partial u_1}{\partial x_3} \\ u_1 \frac{\partial u_2}{\partial x_1} + u_2 \frac{\partial u_2}{\partial x_2} + u_3 \frac{\partial u_2}{\partial x_3} \\ u_1 \frac{\partial u_3}{\partial x_1} + u_2 \frac{\partial u_3}{\partial x_2} + u_3 \frac{\partial u_3}{\partial x_3} \end{bmatrix}$$

or

$$(\mathbf{u} \cdot \nabla \mathbf{u})_i = u_j \frac{\partial u_i}{\partial x_j}.$$

The rank is $1 + (1 + 1) - 2 = 1$.

For the cross product, the curl,

$$\nabla \times \mathbf{u} = \begin{bmatrix} \frac{\partial u_3}{\partial x_2} - \frac{\partial u_2}{\partial x_3} \\ \frac{\partial u_1}{\partial x_3} - \frac{\partial u_3}{\partial x_1} \\ \frac{\partial u_2}{\partial x_1} - \frac{\partial u_1}{\partial x_2} \end{bmatrix}$$

or

$$(\nabla \times \mathbf{u})_i = e_{ijk} \frac{\partial}{\partial x_j} u_k.$$

The rank is $1 + 1 - 1 = 1$.

For the Frobenius inner product

$$\begin{aligned} \boldsymbol{\tau} : \nabla \boldsymbol{U} = {} & \tau_{11} \frac{\partial U_1}{\partial x_1} + \tau_{12} \frac{\partial U_1}{\partial x_2} + \tau_{13} \frac{\partial U_1}{\partial x_3} \\ & + \tau_{21} \frac{\partial U_2}{\partial x_1} + \tau_{22} \frac{\partial U_2}{\partial x_2} + \tau_{23} \frac{\partial U_2}{\partial x_3} \\ & + \tau_{31} \frac{\partial U_3}{\partial x_1} + \tau_{32} \frac{\partial U_3}{\partial x_2} + \tau_{33} \frac{\partial U_3}{\partial x_3} \end{aligned}$$

or

$$\boldsymbol{\tau} : \nabla \boldsymbol{U} = \tau_{ij} \frac{\partial U_i}{\partial x_j}.$$

The rank is $2 + (1 + 1) - 4 = 0$.

References

1. Cercignani, C., *Rarefied Gas Dynamics: From Basic Concepts to Actual Calculations*, 2000, Cambridge: Cambridge University Press.
2. Chhabra, R. P. and J. F. Richardson, *Non-Newtonian Flow and Applied Rheology: Engineering Applications*, 2008, Oxford: Butterworth-Heinemann.
3. Versteeg, H. K. and W. Malalasekera, *An Introduction to Computational Fluid Dynamics: The Finite Volume Method*, 1995, New York: Wiley.
4. Chung, T. J., *Computational Fluid Dynamics*, 2002, New York: Cambridge University Press.
5. Hinze, J. O., *Turbulence*, 1975, New York: McGraw-Hill.
6. Tennekes, H. and J. L. Lumley, *A First Course in Turbulence*, 1972, Cambridge, MA: MIT Press.
7. Pope, S., *Turbulent Flows*, 2000, Cambrige: Cambridge University Press.
8. Bernard, P. S. and J. M. Wallace, *Turbulent Flow*, 2002, Hoboken, NJ: John Wiley & Sons Inc.
9. Garde, R. J., *Turbulent Flow*, 1994, New Delhi: Wiley Eastern Limited.
10. Mathieu, J. and J. Scott, *An Introduction to Turbulent Flow*, 2000, Cambridge: Cambridge University Press.
11. Sagaut, P., *Large Eddy Simulation for Incompressible Flows*, 2001, Heidelberg: Springer.
12. Wilcox, D. C., *Turbulence Modeling for CFD*, 2000, La Cãnada, CA: DCW Industries.
13. Schlichting, H., *Boundary Layer Theory*, 2000, Berlin: Springer.
14. Baldyga, J. and J. R. Bourne, *Turbulent Mixing and Chemical Reactions*, 1999, New York: Wiley.
15. Fox, R. O., *Computational Models for Turbulent Reacting Flows*, 2003, Cambridge: Cambridge University Press.
16. Clift, R., J. R. Grace and M. E. Weber, *Bubbles, Drops, and Particles*, 1978, New York: Academic Press.
17. Crowe, C., M. Sommerfeld and Y. Tsuji, *Multiphase Flows with Droplets and Particles*, 1998, Boca Raton, FL: CRC Press.
18. Kolev, N. I., *Multiphase Flow Dynamics. 1, Fundamentals*, 2005, Berlin: Springer.
19. Gidaspow, D., *Multiphase Flow and Fluidization: Continuum and Kinetic Theory Description*, 1994, Boston, MA: Academic Press.
20. European Research Community on Flow, Turbulence and Combustion (ERCOFTAC), *Special Interest Group on Quality and Trust in Industrial CFD Best Practice Guidelines*, eds. M. Casey and T. Wintergerste, 2000, online.
21. European Research Community on Flow, Turbulence and Combustion (ERCOFTAC), *Special Interest Group on Dispersed Turbulent Multi-Phase Flow. Best Practice Guidelines*, eds. M. Sommerfeld, B. van Wachem and R. Oliemans, 2008, online.

Index

Printed in the United States
By Bookmasters